Global Environment
Problems and Policies

Volume–V

Edited by

K.R. Gupta
Prasenjit Maiti

ATLANTIC
PUBLISHERS & DISTRIBUTORS (P) LTD

Published by

ATLANTIC

PUBLISHERS & DISTRIBUTORS (P) LTD

7/22, Ansari Road, Darya Ganj,
New Delhi-110002
Phones : +91-11-23273880, 23275880, 23280451
Fax : +91-11-23285873
Web : www.atlanticbooks.com
E-mail : info@atlanticbooks.com

Branch Office
5, Nallathambi Street, Wallajah Road,
Chennai-600002
Phones : +91-44-64611085, 32413319
E-mail : chennai@atlanticbooks.com

ISBN 978-81-269-0999-5 Vol. 5
ISBN 978-81-269-0849-3 Set

Printed in India at Nice Printing Press, A-33/3A, Site-IV,
Industrial Area, Sahibabad, Ghaziabad, U.P.

Preface

The 20th century has witnessed major developments, which have momentous impact not only on the lives of human beings but also on planet earth—the only known planet in our solar system that supports life by virtue of its unique environmental conditions—water, an oxygen-rich atmosphere, and a suitable surface temperature. Limitless greed, reckless consumption of natural resources and unkind treatment meted out to environment have increasingly damaged the earth. An unprecedented growth in population and consumption, rapidly increasing urbanization, dramatic changes in the global economic system and the revolution in communication technology have created several problems for society. Now there is an increasing awareness that the process of development including industrialization and economic growth has come at a huge cost to society, to the environment and to our very future.

It is ironical that some of the achievements of the 20th century themselves pose the challenges for the present 21st century. As of 2008, the world population has reached 6.71 billion. According to data from the CIA's *World Fact Book*, world human population increases by 211, 090 people every day. With 6.71 billion people inhabiting this earth, the global economy over the next 25 years could expand from $35 trillion in 2005 to $72 trillion in 2030 (World Bank). The pace of human and economic activity is putting pressure on global life support systems.

Today, we witness the depletion of non-renewable natural resources, the continued destruction of life sustaining forests and illegal international trade in protected species of flora and fauna. There is an ongoing depletion of our planet's biological treasures and the threat of acceleration in global warming and in the greenhouse effect. While the behaviour of monsoon is changing sharply, the glaciers are receding because of global warming. The total surface area of glaciers worldwide has decreased 50 per cent since the end of the 19th century. The retreat has been stark in the Himalayas. It may be noted here that the Himalayan glaciers are sensitive indices of climate change due to their typical vulnerability; snow accumulation and melting on these depends on monsoonal precipitation as well as summer temperatures. To quote from the Fourth Assessment Report released by IPCC in 2007, "It is likely that future topical cyclones (typhoons and hurricanes) will become more intense, with larger peak wind speeds and more heavy precipitation associated with ongoing increases of sea surface temperatures." Thus, it is likely that the increase in atmospheric temperature resulting in increase in precipitation on glaciers will eventually affect a variety of planning issues such as the planning and design of hydrological structures, river basin management, flood control and drought management, urban planning and industrial development and above all, agricultural policies. To put it simply, climate change is a larger challenge of sustainable development that affects global economic growth.

Besides, there is a rapid globalisation of environmental destruction. The earth is fast losing its treasures. The soil we cultivate, the water we drink and the air we breathe are all polluted. The environmental impact of human activity is no longer restricted to man-made national boundaries.

There is a paramount need to create widespread consciousness of the environment. It must permeate all ages and all sections of society, beginning with the child. The environmental problems call for more than scientific and technological solutions. They require the re-establishment of environmental ethics and values practised in traditional societies, with economic development based on these issues.

For developing awareness among the readers about the preservation of biodiversity and its significance for life, sustainability and equity, and for warning against the consequences of environmental exploitation as well, the present volumes have been compiled. These have a wide coverage and comprehensively study not only the aforesaid problems but also the policies and schemes relating to the environment.

A major impetus in this regard was provided by the 1987 Report of the World Commission on the Environment and Development (the Brundtland Report) and was accelerated by the Rio Summit in 1992. Although the environmental impact assessment was first formally established in USA in 1969, the new techniques and approaches, policies and agreements have been continuously evolved to prevent damage to existing resources and to promote its conservation and augmentation. In this direction, much headway was made in the 12th Conference of Parties to the UN Climate Convention held in Nairobi, Kenya in November 2006, in which it was decided to set up an adaptation fund to help poor countries adapt to climate change.

The inaugural session of the Delhi Sustainable Development Summit was held in February 2008 in the background of the increasing concerns related to climate change in the world today. The summit assumes greater significance as it connects issues related to sustainable development with climate change and hopes to encourage activities that need to be undertaken for sustainable development across the globe. The 28th session of the Subsidiary Body for Scientific and Technological Advice (SBSTA) and the Subsidiary Body for Implementation (SBI) of the United Nations Framework Convention on Climate Change took place in Bonn in June 2008 in which various pertinent issues were discussed. The Ad Hoc Working Group on Kyoto Protocol is also set to complete its work by the end of 2009. It is hoped that all these efforts being made to combat environmental pollution and climate change would translate into sustainable development.

The book would be highly useful to students and researchers on Environment. It would also be useful to Government executives and NGOs concerned with environment and pollution problems. Since the laymen are deeply interested in clean environment, the book would be of great interest to them.

K.R. Gupta
Prasenjit Maiti

Contents

Sustainable Development: A Key Environmental Debate

1

B. Anjan Kumar Prusty and *P.A. Azeez*

Introduction

One of the leading focuses within the environmental debate during the 1990s/end of last century has been the search for more sustainable ways of using the earth and its resources. Many symptoms of the environmental crisis illustrate what can go wrong if environmental resources are over used or used in ways that create environmental damage or instability. This is particularly true for Lower Developed Countries (LDCs) where, More Developed Countries (MDCs) notions of development have often been transplanted with dire consequences. A central objective of the 1992 Rio Earth Summit was to advocate the need to replace existing exploitative and environmentally damaging forms of economic development with more sustainable and environmentally friendly forms of development. The term sustainable development was made popular through the Brundtland Report, which called for a change in the economic world order to reduce the destruction of the environment and solve social ills (Golley et al., 1994). The Brundtland Report (report of The World Commission on Environment and Development) received widespread support and approbation and the term sustainable development has become a code phrase among the international agencies, scholars and practitioners working in the global arena. Because of its popularity the term has been defined in many ways to fit varied purposes.

The themes of integrating environmental concerns into all aspects of development programmes, of sustainable technology transfer and of trade and aid patterns between MDCs and LDCs are now being more widely discussed. Familiar themes such as continued population growth, widespread poverty and gaining control over natural resource depletion are being revitalized within the emergent sustainable development framework. Recognition of the need to protect the resource rights and interests of future generations is fostering interest in intergenerational equity as a key ingredient of sustainability thinking. The sustainability debate is lively at a variety of spatial scales, ranging from the global to the local. The emergence of this (sustainable development) theme has also been accompanied by a renewed interest in world futures.

Definition

At the first glance the word sustainable development seems to be an oxymoron, conjoining two contradictory terms, 'sustain' and 'development'. Development means to expand through growth, to advance, to evolve. Sustain means to maintain, continue, keep up, to prolong. Thus sustainable development might be defined as prolonged or continuous growth. However, the broader concept of sustainable development was first introduced during late 1980s, by the World Commission on Environment and Development. In its 1987 report *Our Common Future* (commonly known as the Brundtland Report) the WECD (1987) defined sustainable development as 'meeting the needs of the present without compromising the ability of the future generations to meet their needs', and this definition has since been widely adopted. This report focused on the global human community and the global environment. At that level sustainable development can mean managing human populations and the global environment so that human life on the planet can continue to evolve and change, while at the same time the health of the natural environment is maintained. Sustainable development thus means extending progress, without exhausting resources, beyond the foreseeable future.

In a world of rapid technological change, what is needed is a concept of sustainability that looks forward, not nostalgically backward at indigenous civilizations and traditions. A workable definition of sustainability must allow for continuing rapid change. Therefore, efforts at achieving sustainability must focus on technological change, seeking commerce and industry more efficient while using science to devise better ways of measuring our progress (Schwab, 1997). Sustainability has become a central theme of environment and development and resource use. Although the idea of sustainability has many facets, the central idea is that we should use resources in ways that don't diminish them. Resources and natural amenities, including wildlife, natural beauty, and open space should be preserved so that future generations can have lifestyles at least healthy and happy as ours or perhaps better (Cunningham and Cunningham, 2002).

Conservationists and ecologists have long been aware of the significance of sustainability within natural environmental systems. The search for a single definition of 'sustainability' seems elusive partly because it embodies a number of ideas imported from different disciplines—including economics (no growth or slow growth), ecology (integrity of the biosphere, carrying capacity), sociology (critique of technology) and environmental studies (ecodevelopment, resource-environment links, environmental ethics). Whilst the general definition has been widely accepted, there has been widespread debate about what sustainability might actually mean in practice, and about how it might best be applied to different cultures and economics. It is generally accepted that development can be sustainable only if it is based on sound ecological principles and practices, but beyond that there is little real consensus. So far too much attention has been focused on defining the meaning of sustainability and not enough on exploring the implications of sustainability, as it is likely to affect the status quo within and between countries.

The Context of Sustainable Development

It has long been recognised that the future of the planet and its people is intimately tied up with our ability to maintain and preserve the life-support systems that nature provides. This makes it our duty to ensure that:

- All uses of renewable resources are sustainable

- The diversity of life on earth is conserved
- Damage to natural environmental systems is minimised.

Sustainability is an in-built feature of all natural environmental systems provided that human interference is absent or minimised. It relates to the capacity of a system to maintain a continuous flow of whatever each part of that system needs for a healthy existence. Human use of environmental resources and interference with environmental systems disturbs this inbuilt capacity, which can make it unsustainable (Viederman, 1993). Traditional ways of harvesting natural renewable resources—such as fish from the oceans, wood from the forest, and plants and products from natural ecosystems—have usually been sustainable so long as the quantities extracted were not too large. Sustainable yield means taking no more from an ecosystem than it can create and at the same time remain healthy, diverse and self-perpetuating. Many types of natural resource uses are now intensive and unsustainable, without deliberate management intervention to limit the quantities removed to below threshold sustainable yields. For sustainable development harvesting renewable resources should be within regeneration rates, waste emissions should not exceed the assimilative capacity of the ecosystem and exploitation of non-renewable resources should be within creation of substitutes (Azeez, 2004).

Implications

One significant effect of the sustainable development debate has been to raise awareness of the need to adopt wider perspectives when making decisions that affect the environment. The (three) issues (within sustainable development) that are now taken much more seriously are (Park, 1997; Day, 1996-1997):

- Inter-generational implications of patterns of resources use: how effectively do decisions about the use of natural resources preserve an environmental heritage or estate for the benefit of future generations?
- Equity concerns: who has access to resources? How fairly are available resources allocated between competing claimants?

- Time horizons: how much are resource allocation decisions oriented towards short-term economic gain or long-term environmental stability?

There is a need of some radical changes in attitudes, values and practices while stepping towards a more sustainable way of living. Moreover we are not sure whether we can create a vibrant world economy that does not destroy the ecosystem on which it is based. The environmentally sustainable 'brave new world' will have to be:

- less polluting, probably heavily reliant on perpetual as well as renewable sources of energy like solar and wind energy
- make extensive use of new ways of using and reusing materials
- adopt less resource-intensive means of growing food and
- develop effective strategies for preserving forests.

There is also a need in the change in people's attitude of resource use, i.e., form 'throw away approach to resource reuse and recovery approach', so that the 'no waste/no waste in the world' target can fully be achieved. There would also need to be radical changes in energy systems, tax systems, international economic structures and provisional of international aid (Miller, 1996). There would also need to be strategic eradication of poverty (the most critical threat to sustainable development), which is widening the gap between the poor and the rich by the day (GEF, 2002).

Principles of Sustainable Development

Sustainable development is a dynamic process, which enables all people to realise their potential and improve their quality of life in ways that simultaneously protect and enhance the Earth's life support systems (Viederman, 1993). But the difficult part is to reconcile the moral obligation to today's poor without compromising what we owe our future generations. A more difficult aspect is compromising the luxury needs of the rich. The concept of sustainable development is based on the triple bottom line connecting economic, social and environmental factors. The strategy for sustainable development is based on nine key principles (Park, 1997), the first of which provides an ethical base, the next four definite criteria that should be met and the last four definite directions that should be adopted are as follows:

- respect and care for the community of life
- improve the quality of human life
- conserve the Earth's vitality and diversity
- minimise the depletion of non-renewable resources
- keep within the Earth's carrying capacity
- change personal attitudes and practices
- enable communities to care for their own environments
- provide a national framework for integrating development and conservation
- create a global alliance.

Development of a global commitment to sustainable development is an urgent need. But radical changes in ethical and cultural policies and principles are the most important to achieve sustainability, as it is an ethical guiding principle, not something to be defined, but to be declared. Retaining the ethical and moral underpinnings of sustainability is absolutely essential to pursue the path of sustainable development.

There are cases where one suspects that the word 'sustainable' is used carelessly, perhaps as though the belief exists that the frequent use of the adjective "sustainable" is all that is needed to create a sustainable society (Bartlett, 1997-98). Sustainable society, a core issue within the debate of sustainable development, will differ from our present society and propel the world to achieve sustainability. Life in a viable ecosystem must become the core value of a sustainable society; that means all life, not just human life (Milbrath, 1996). We must give top priority to the ecosystems that support us, and second priority to our societies. A sustainable society would:

- affirm love as a primary value and extend it to future generations and other species

- emphasize partnership rather than domination, cooperation over competition
- focus on earth and life-centred environmental worldviews rather than human centred worldviews
- encourage self-realization—as the key to a fulfilling life
- follow the principles of long-term development with emphasis on environmental stability rather than Cornucopian (short-term economic development) path
- continue further development of science and technology because we need practical creative solutions that are both environmentally sound and economically feasible.

Conclusion

The continued increase in numbers of humans and expansion of human demands upon the natural environment cannot continue indefinitely. The problem of maldistribution of material life, which far exceeds the geographic distribution of resources, must be solved. Thus sustainable development will result in the replacement of the present society by a more sustainable one, which will ensure the health and vitality of human life and culture and of nature, for present and future generations, by eradicating the poverty, thus narrowing the gap between the poor and rich; by ending the activities that destroy human life and culture and nature; by conserving what exists, restoring what has been damaged and preventing future harm.

WORKS CITED

Azeez, P.A. (2004). "Sustainable Development and Environmental Impact Assessment." *Proceedings of The National Workshop on Recent Trends in Pollution Control and Environmental Conservation* (Ed. R Sivakumar), Bannari Amman Institute of Technology, Sathyamangalam, Tamil Nadu, India.

Bartlett, A.A. (1997-98). "Reflections on Sustainability, Population Growth and the Environment-Revisited." *Renewable Resources Journal.* 15 (4): 6-23.

Cunningham, W.P. and Cunningham, M.A. (2002). *Principles of Environmental Science: Inquiry and Applications.* McGraw Hill Publications.

Day, R.D. (1996-97). "Future Directions for Sustainable Development and Resource Sustainability." *Renewable Resources Journal.* 14 (4): 12-14.

Global Environment Facility (GEF, 2002). "The Challenge of Sustainability: An Action Agenda for the Global Environment." Washington, DC.

Golley, F., Baudry, J., Berry, R.J., Bornkamm, R., Dahlberg, K., Jansson, A.M., King, J., Lee, J., Lenz, R., Sharitz, R.J., Svedin, A.U. (1994). "What is the Road to Sustainability?" *Renewable Resources Journal.* 12 (2): 12-15.

Milbrath, L.W. (1996). "Envisioning a Sustainable Society (Guest Essay)." *Living in The Environment.* (G. Tyler Miller, Jr.), Wadsworth Publishing Company, New York, 724-25.

Miller, G.T. (1996). *Living in The Environment.* Wadsworth Publishing Company, New York, 710-25.

Park, C. (1997). *The Environment: Principles and Applications.* Routledge, London & New York, 23-27.

Schwab, J. (1997). "Sustainable Development: Converting Motherhood and Apple Pie into Substance." *Renewable Resources Journal.* 15 (2): 16-19.

Viederman, S. (1993). "A Dream of Sustainability." *Renewable Resources Journal.* 11 (2): 14-15.

World Commission on Environment and Development (1987). *Our Common Future: Report of the World Commission on Environment and Development.* Oxford University Press, New York.

Hydrogen Economy might Impact Earth's Stratosphere: A Survey of Literature

2

Biswajit Mandal

Foreword

Energy is the very lifeblood of today's society and economy. Our work, leisure, and our economic, social and physical welfare all depend on the sufficient, uninterrupted supply of energy. Yet we take it for granted—and energy demand continues to grow, year after year. Traditional fossil energy sources such as oil are ultimately limited and the growing gap between increasing demand and shrinking supply will, in the not too distant future, have to be met increasingly from alternative primary energy sources. We must strive to make these more sustainable to avoid the negative impacts of global climate change, the growing risk of supply disruptions, price volatility and air pollution that are associated with today's energy systems. The energy policy of the European Commission advocates securing energy supply while at the same time reducing emissions that are associated with climate change. This calls for immediate actions to promote greenhouse gas emissions-free energy sources such as renewable energy sources, alternative fuels for transport and to increase energy efficiency.

On the technology front, hydrogen, a clean energy carrier that can be produced from any primary energy source, and fuel cells which are very efficient energy conversion devices, are attracting the attention of public and private authorities. Hydrogen and fuel cells, by enabling the so-called hydrogen economy, hold great promise for meeting in a quite unique way, our concerns over security of supply and climate change.

With these factors in mind, we established the High Level Group for Hydrogen and Fuel Cell Technologies in October 2002, and asked its members to come forward in six months with a collective vision of how these technologies could help meet Europe's aspirations for sustainable energy systems. This report is the result and, we believe, a first milestone.

The report highlights the need for strategic planning and increased effort on research, development and deployment of hydrogen and fuel cell technologies. It also makes wideranging recommendations for a more structured approach to European Energy policy and research, for education and training, and for developing political and public awareness. Foremost amongst its recommendations is the establishment of a European Hydrogen and Fuel Cell Technology Partnership and Advisory Council to guide the process.

Security of energy supply is of major concern for the European Union. As North Sea production peaks, our dependence on imported oil—vital for today's transport systems—is forecast to grow from around 75 per cent today, to in excess of 85 per cent by 2020, much of it coming from the Middle East. We have also witnessed the disruption and economic loss caused by recent major grid outages in North America and Italy, illustrating the need to reinforce security of supply. In the transatlantic summit held on 25th June 2003 in Washington, President Prodi, Prime Minister Simitis and President Bush stated that the European Union and the United States should co-operate to accelerate the development of the hydrogen economy as a means of addressing energy security and environmental concerns.

Hydrogen based energy systems can build bridges to the future, but planning a cost-effective and efficient transition is hugely complex. The very large capital and human investments implied will require many years before coming to fruition. However, we must begin now to explore this path to a more sustainable future.

The High Level Group's vision was presented at the conference "The hydrogen economy—a bridge to sustainable energy" held in Brussels in June 2003 and presided over by

President Prodi. The group's vision and recommendations were strongly supported. We, therefore, endorse the recommendations of the High Level Group and the need for action today. That is why we intend to launch a "European Partnership for the Sustainable Hydrogen Economy" as soon as possible, to mobilize a broad range of stakeholders and structure a coherent effort on advancing sustainable hydrogen and fuel cell technologies in Europe.

Finally, we wish to thank the members of the High Level Group and their "sherpas" for the very considerable time and effort put in to reaching this collective vision, which we believe will prove influential in paving the way to a sustainable hydrogen economy.

Hydrogen as a Fuel

Hydrogen is the lightest (~6.1 × 10-3 lb/ft^3) and one of the most abundant elements in the universe. However, it almost always occurs in combined form as in water (H_2O), methane (CH_4), or coal, which contains 2-6 per cent hydrogen by weight. Even when it is separated from other elements, it is stable only when two hydrogen atoms are combined into a single molecule (H_2). Its energy density, due to an electro-chemical reaction or combustion with oxygen on a per-unit-mass basis, is quite high (~51,600 Btu/lb; lower heating value)1, but on a per-unit-volume basis, it is among the lowest of any fuels (~315 Btu/ft^3 at atmospheric pressure). The energy density on a per-unit-volume basis can be increased by compressing gaseous hydrogen to high pressures (5,000 to 10,000 psi) or by liquefying it at cryogenic temperatures (about 20°K at atmospheric pressure).

However, both 10,000 psi gaseous hydrogen and cryogenic liquid hydrogen have volumetric energy densities considerably lower than gasoline and diesel fuel. These properties make hydrogen very difficult to store, particularly on transportation vehicles, and difficult to distribute from one location to another. Hydrogen also burns with an almost invisible flame and is subject to special handling regulations by state and national codes for safety reasons. Clearly, the use of hydrogen as a transportation fuel has many engineering challenges that may be difficult to address. It is well known that hydrogen, like electricity, is an energy carrier, not an energy resource. It does not occur naturally in nature and must be extracted from other sources such as water, hydrocarbon fuels, or other naturally occurring materials containing hydrogen. This extraction requires energy from other sources, i.e., solar energy (hydro, wind, or photovoltaic), fossil-fuel combustion, nuclear fission, or combustion of renewable materials (wood, fibrous crops, bagasse, or even garbage). In every case, the energy content of hydrogen is less, often significantly less, than the energy it takes to produce it. This is also true of electricity because its energy is only 30 per cent to 40 per cent of the thermal energy used to produce it. The energy content of hydrogen varies from 20 per cent to 80 per cent of the thermal energy used to produce it, depending upon the method used and the source (i.e., water, methane, coal, etc.). Hence, both hydrogen and electricity are premium energy carriers whose use can be justified only by special benefits associated with their use. In the case of hydrogen, a special benefit is that it can be converted into electricity for transportation using fuel cells with an efficiency that is at least twice as high as the conversion in thermodynamic heat engines. Other benefits include the reduction of pollutants and greenhouse gases being emitted into the atmosphere by transportation vehicles using hydrogen as a fuel. A recent document indicates that "For transportation propulsion applications, DOE is focusing on direct hydrogen fuel cells, in which on-board storage of hydrogen is supplied by a hydrogen generation, delivery, and fueling infrastructure" *[DOE, 2003]*. The essential components of a hydrogen economy are production of hydrogen, storage of hydrogen, distribution or transportation of hydrogen to vehicle fuel stations, dispensing of hydrogen into vehicles, and use of hydrogen in fuel-efficient engines (i.e., fuel cells rather than either internal- or external-combustion engines). Each of these activities constitutes an engineering challenge that will be discussed here.

Hydrogen vs Liquid Hydrocarbons

In many ways, hydrogen is a good deal safer than gasoline or diesel. Like any fuel, hydrogen stores significant amounts of energy, and handling it requires certain safety precautions. But hydrogen can be safer than gasoline if it is used properly.

Because it is so light, hydrogen disperses and floats skyward when leaked—it won't pool or soak into clothing like gasoline, just waiting to ignite. (Spilled hydrogen won't soak into the earth and pollute ground water either, or cause an environmental disaster like the Exxon Valdez.)

But what if the hydrogen does somehow ignite in a car? Tests conducted at the College of Engineering at Miami University aimed to find this out. 3000 cubic feet per minute of hydrogen was leaked from a vehicle tank and set alight. Over the course of the burn, temperature sensors inside the vehicle did not measure an increase of more than 1 or 2 degrees centigrade anywhere inside the vehicle. The temperature of the surface of the outside of the vehicle did not climb above that of a vehicle sitting in the sunshine!

This might sound unintuitive. But when a carbon-based fuel like gasoline burns, glowing hot soot particles transfer the heat to its surroundings—potentially including you. But because hydrogen contains no carbon, it burns cleanly without a residue of hot soot, producing little radiant energy. This means that a victim would have to be practically in the flame in order to get burned.

Pressurized hydrogen tanks are made to withstand enormous impacts, and fail gracefully, if at all. Some fear that a hydrogen tank has the potential to explode, and that is possible. But these critics often overlook the greater explosive potential of the gas tanks in their vary own cars.

Many real-life tests have demonstrated the safety of pressurized hydrogen storage. Simulated 55 mph crash tests left the car totaled, but the hydrogen tank intact. To prove the safety of its hydrogen vehicles, BMW tested its hydrogen tanks in a series of accident simulations that included collision, fire and tank ruptures. In all cases, the hydrogen cars fared as well as conventional gasoline vehicles. And hydrogen-fueled cars are designed to preclude the possibility of leaked hydrogen collecting within the vehicle.

The *Hindenburg* Myth

Most hydrogen concerns stem from the Hindenburg disaster of 1937. The hydrogen gas that once filled the Hindenburg zeppelin did burn, but it did so quickly, upwardly, and away from the people below. When the airship was docking, an unexpected electrical discharge ignited the airship's canvas (which was unknowingly treated with two major components of rocket fuel!). The clean hydrogen flames swirled above the occupants of the passenger compartment, and all those who rode the airship down to the ground survived. 35 of the 37 casualties perished from jumping to the ground, and most other injuries resulted from diesel burns.

Impacts on Earth's Stratosphere

According to conventional wisdom, hydrogen-fueled cars are environmentally friendly because they emit only water vapor—a naturally abundant atmospheric gas. But leakage of the hydrogen gas that can fuel such cars could cause problems for the upper atmosphere, new research shows.

In an article appearing in the journal Science, researchers from the California Institute of Technology report that the leaked hydrogen gas that would inevitably result from a hydrogen economy, if it accumulates, could indirectly cause as much as a 10 per cent decrease in atmospheric ozone. The researchers are physics research scientist Tracey Tromp, Assistant Professor of Geochemistry John Eiler, planetary science professor Yuk Yung, planetary science research scientist Run-Lie Shia, and Jet Propulsion Laboratory scientist Mark Allen.

If hydrogen were to replace fossil fuel entirely, the researchers estimate that 60 to 120 trillion grams of hydrogen would be released each year into the atmosphere, assuming a 10 to 20 per cent loss rate due to leakage. This is four to eight times as much hydrogen as is currently released into the atmosphere by human activity, and would result in doubling or tripling of inputs to the atmosphere from all sources, natural or human. Because molecular hydrogen freely moves up and mixes with stratospheric air, the result would be the creation of additional water at high altitudes and, consequently, an increased dampening of the stratosphere. This in turn would result in cooling of the lower stratosphere and disturbance of ozone chemistry, which depends on a chain of chemical reactions involving hydrochloric acid and chlorine nitrate on water ice.

The estimates of potential damage to stratospheric ozone levels are based on an atmospheric modeling program that tests the various scenarios that might result, depending on how much hydrogen ends up in the stratosphere from all sources, both natural and anthropogenic.

Ideally, a hydrogen fuel-cell vehicle has no environmental impact. Energy is produced by combining hydrogen with oxygen pulled from the atmosphere, and the tailpipe emission is water. The hydrogen fuel could come from a number of sources (Iceland recently started pulling it out of the ground). Nuclear power could be used to generate the electricity needed to split water, and in principle, the electricity needed could also be derived from renewable sources such as solar of wind power.

By comparison, the internal combustion engine uses fossil fuels and produces many pollutants, including soot, noxious nitrogen and sulfur gases, and the "greenhouse gas" carbon dioxide. While a hydrogen fuel-cell economy would almost certainly improve urban air quality, it has the potential unexpected consequences due to the inevitable leakage of hydrogen from cars, hydrogen production facilities, and the transportation of the fuel.

Uncertainty remains about the effects on the atmosphere because scientists still have a limited understanding of the hydrogen cycle. At present, it seems likely such emissions could accumulate in the air. Such a build-up would have several consequences, chief of which would be a moistening and cooling of the upper atmosphere and, indirectly, destruction of ozone.

In this respect, hydrogen would be similar to the chlorofluorocarbons (once the standard substance used for air conditioning and refrigeration), which were intended to be contained within their devices, but which in practice leaked into the atmosphere and attacked the stratospheric ozone layer.

The authors of the Science article say that the current situation is unique in that society has the opportunity to understand the potential environmental impact well ahead of the growth of a hydrogen economy. This contrasts with the cases of atmospheric carbon dioxide, methyl bromide, CFCs, and lead, all of which were released into the environment by humans long before their consequences were understood.

"We have an unprecedented opportunity this time to understand what we're getting into before we even switch to the new technology," says Tromp, the lead author. "It won't be like the case with the internal-combustion engine, when we started learning the effects of carbon dioxide decades later."

The question of whether or not hydrogen is bad for the environment hinges on whether the planet has the ability to consume excess anthropogenic hydrogen, explains Eiler. "This man-made hydrogen will either be absorbed in the soil—a process that is still poorly understood but likely free of environmental consequences—or react with other compounds in the atmosphere.

"The balance of these two processes will be key to the outcome," says Eiler. "If soils dominate, a hydrogen economy might have little effect on the environment. But if the atmosphere is the big player, the stratospheric cooling and destruction of ozone modeled in this Science paper are more likely to occur.

"Determining which of these two processes dominates should be a solvable problem," states Eiler, whose research group is currently exploring the natural budget of hydrogen using new isotopic techniques.

"Understanding the effects of hydrogen on the environment now should help direct the technologies that will be the basis of a hydrogen economy," Tromp adds. "If hydrogen emissions present an environmental hazard, then recognizing that hazard now can help guide investments in technologies to favor designs that minimize leakage.

"On the other hand, if hydrogen is shown to be environmentally friendly in every respect, then designers could pursue the most cost-effective technologies and potentially save billions in needless safeguards."

"Either way, it's good for society that we have an emission scenario at this stage," says Eiler. "In past cases—with chlorofluorocarbons, nitrogen oxides, methane, methyl bromide, carbon dioxide, and carbon monoxide—we always found out that there were problems long after they were in common use. But this time, we have a unique opportunity to study the anthropogenic implications of a new technology before it's even a problem."

If hydrogen indeed turns out to be bad for the ozone layer, should the transition to hydrogen-fueled cars be abandoned? Not necessarily, Tromp and Eiler claim.

"If it's the best way to provide a new energy source for our needs, then we can, and probably should, do it," Tromp says.

Eiler adds, "If we had had perfect foreknowledge of the effects of carbon dioxide a hundred years ago, would we have abandoned the internal combustion engine? Probably not. But we might have begun the process of controlling CO_2 emissions earlier."

Hydrogen is too Dangerous

One major reason that the US has been so slow to develop a hydrogen economy is the public perception that hydrogen is too dangerous for common use. Actually, in regard to safety, hydrogen is very much like propane, which is widely used. Both gases require careful handling and respect, but they can be safely used for every day applications.

It can be argued that hydrogen is safer than gasoline. Leaking hydrogen is quite unlikely to explode because it dissipates upward very quickly, and it takes four times as high a concentration of hydrogen as of gasoline vapor to ignite in the air.

The Hindenburg dirigible disaster in 1937 gave hydrogen a big public relations problem. However, it has recently been determined that the Hindenburg was covered with cloth doped with aluminum and iron oxide, now known to be a highly explosive and unstable substance (later used for rocket fuel!). It was the cloth doping which ignited from atmospheric electricity. In turn, this ignited the hydrogen inside, which of course contributed to the conflagration.

Unfortunately, the huge and terrible spectacle captured on film has made an indelible impression on the public psyche, juxtaposing the idea of hydrogen fuel with the Hindenburg disaster. This created the extremely widespread idea, continuing even into the 21st century, that hydrogen is too dangerous for domestic use as a fuel. This mass misperception crippled development of hydrogen technology for consumers.

Hydrogen is too Expensive

Well, at the moment it is too expensive for most of us, but that's because hydrogen development is at a "vicious circle" stage where there is not much economic incentive to make hydrogen fuel and sell it because consumer applications (hydrogen vehicles and appliances) are not widely available and affordable; and there is not much economic incentive to manufacture hydrogen appliances and vehicles because the fuel is not widely available and affordable.

So, right now, in the spring of 2005, hydrogen fuel for consumers is a novelty. However, this is not because hydrogen technology is complex and we don't have the capability to work with it, or that it is inherently expensive to produce hydrogen fuel.

On the contrary, for example, a competent do-it-yourself can build a hydrogen production system that uses renewable energy sources, and also build fuel cells to generate electricity from the hydrogen. Appliances that burn propane or natural gas can be converted to hydrogen reasonably easily.

The irony is that hydrogen is already in widespread use in the United States in industry, including refining gasoline and diesel. The oil companies certainly wouldn't be using it if it wasn't cost effective for them.

A very large portion of the price consumers pay for fuel is for the cost of transporting it—in the case of oil and gasoline, from the well-head, often continents away, to our home town. Because hydrogen is so light, transporting it, especially over long distances, requires slightly different strategies that can involve added energy to compress and store it. However, this consideration is not as relevant as it might seem.

When hydrogen's cost to consumers is evaluated what is often missing from the equation is the fact that hydrogen can be made nearly anywhere, from any power source, including renewable energy sources—so to begin with, most or even all transportation costs can be eliminated. Why on earth should we pay to transport fuel thousands and thousands of miles when it can be made locally or even in our backyards? Per the US Department of Energy's figures, the present price of delivered liquid hydrogen is around four times the cost of producing the hydrogen. That is a lot of financial incentive to produce it locally!

What the oil companies don't like about hydrogen is this previous-mentioned fact, and there is not much about it in the mass media. It will be very difficult for energy monopolies to survive in the coming hydrogen economy because a centralized infrastructure like the present one for fossil fuel will be superfluous, and people will wake up to this fact sooner or later, both as entrepreneurs and consumers. Renewable energy resources cannot be controlled in any way comparable to the way control is exercised over fossil fuel resources by a relatively small number of people. It will be not be so easy for a very few to become obscenely wealthy in the hydrogen business.

If you wish to truly compare the cost of hydrogen to fossil fuels, you must also consider the hidden costs of the fossil fuel economy. What "price" for the damage and destruction done to the health of people, ecosystems and economies, and who pays? The horror of huge disasters like the oil tanker wrecks of the Exxon Valdez in Alaska, or the Prestige off Spain's Galician coast illustrate this issue dramatically, but even business as usual in the oil economy is an assault on the planet and all its inhabitants—we just don't find the little trickles of pollution all around us to be so newsworthy. The quality of our air, water and soil continues to be seriously compromised by the burning of fossil fuels and day-to-day "minor" spills and accidents here and there—some not so minor, like the routine dumping of jet fuel by both commercial and military aircraft. (It's done over both land and sea, by the way.)

So, compare this to hydrogen. For hydrogen being transported, the result of leaking and spills is...water. When it is "burned", the exhaust is...water.

Finally, in any cost comparison of hydrogen to other fuels, be sure that you're not looking at a comparison of apples to oranges. It's not really meaningful to compare the price of a gallon of hydrogen to a gallon of gasoline. What really counts is how many cents a mile your fuel costs, or how many cents per BTU it takes to heat your house. Using Amory Lovins' figures, even at the present price of delivered liquid hydrogen, if you used it to power a hydrogen fuel cell vehicle, your cost per mile would be about like getting gasoline for a dollar a gallon. It is reasonable to expect that as the hydrogen infrastructure is developed, the cost could be significantly lower, and if you make it yourself it could be much lower.

Hydrogen is not a Clean Fuel

Hydrogen as fuel creates no toxic exhaust, but it is only really as clean as the energy source it is derived from. Obviously, producing hydrogen from fossil fuels creates pollution, though looking at the entire process including fuel production, transport and consumption; there is an overall reduction in pollutants created per mile driven, if you compare it to gasoline and diesel in vehicle applications, for instance. It is also potentially easier to control the pollution created by using hydrogen made from "dirty" sources, like coal and oil, because the pollution is limited to the fuel production process, instead of occurring everywhere from the wellhead to your vehicle's tailpipe.

Any such "dirty" hydrogen should be viewed as transitional to hydrogen produced from non-polluting renewable energy sources. No doubt the oil companies will want to drag out the transition as long as possible.

Hopefully, this transition will be kept in mind as we move into a hydrogen economy. Ideally, we should skip as much as we can of the "dirty" hydrogen phase, and really devote our resources right away to producing clean hydrogen. It can certainly be done...but who is going to do it?

It's not Efficient to Produce Hydrogen from Renewable Energy Sources

How do you measure efficiency? What may not be efficient on a large scale may be quite efficient on a smaller scale. 20th century thinking went toward bigger is more efficient, a la Henry Ford's assembly lines. We now know that is certainly not always the case.

Again, one of the major issues of cost and efficiency is the transporting of fuel. About two-thirds of the cost of a gallon of gasoline is the cost of transporting it to the pump from the well-head. Locally produced hydrogen gets points for efficiency just because of the elimination of most of the transporting.

Why take electricity from solar panels or wind generators and make hydrogen to make electricity? Well, it would be great if you could just plug in directly to your solar panel or wind generator, but the fact is, you must have some kind of storage to have a reliable system that provides energy whenever you need it—whether or not the sun's out or the wind's blowing. Hydrogen in this context can store energy that would otherwise go to waste. Also, the use of hydrogen generated by renewables makes it possible to fuel vehicles from solar, wind and hydro power, so a household, business or community could potentially provide for all its energy needs on site or locally.

Solar technology is now progressing fast, after being largely neglected for many years. For instance, roofing material coated with a durable photovoltaic film has recently become available to builders. Simple innovations in solar panel design are significantly increasing the efficiency of electrolysis via photovoltaics, so that even on cloudy days there can be good hydrogen production. As the demand for solar power increases, the price of commercially available panels will become more affordable. Photovoltaics may still be a novelty for most people, but not for long. It will soon become ubiquitous as the technology matures, and using hydrogen to store solar energy will also become commonplace.

Hydrogen Storage

Although molecular hydrogen has excellent energy density on a mass basis, as a gas at ambient conditions it has poor energy density per volume. As a result, if it is to be stored and used as fuel onboard the vehicle, molecular hydrogen must be pressurized or liquefied to provide sufficient driving range. Increasing gas pressure improves the energy density per volume, making for smaller, but not lighter container tanks. Achieving higher pressures necessitates greater use of external energy to power the compression. Alternatively, higher volumetric energy density liquid hydrogen may be used. However liquid hydrogen is cryogenic and boils at 20.268 K (–252.882°C or –423.188°F). Cryogenic storage cuts weight but requires large liquification energies. The liquefaction process, involving pressurizing and cooling steps, is energy intensive. The liquefied hydrogen has lower energy density per volume than gasoline by approximately a factor of four. Storage tanks must also be well insulated to minimize boil off. Ice may form around the tank and help corrode it further if the insulation fails. Insulation for liquid hydrogen tanks is usually expensive and delicate.

The mass of the tanks needed for compressed hydrogen reduces the fuel economy of the vehicle. Because it is a small energetic molecule, hydrogen tends to diffuse through any liner material intended to contain it, leading to the embrittlement, or weakening of its container. This is called *the storage problem*.

The most common method of onboard hydrogen storage in today's demonstration vehicles is as a compressed gas at pressures of roughly 700 bar (70 MPa). Many people believe that the energy needed to compress hydrogen to these pressures presents a major barrier to a hydrogen economy. For example, if one considers the entire world using hydrogen just in their cars, then a massive amount of energy would be needed simply to compress the hydrogen for storage. If this energy was not recovered in any way, the net energy used to compress it would be wasted. Currently, vehicle fuel cells are very expensive, typically 100 times more expensive per kW output than conventional internal combustion engines. It further has been suggested that cars utilizing Li-ion or Li-polymer batteries for onboard energy storage are capable of being more efficient than hydrogen-fueled cars would ever be, and that they just need to be mass produced to become cost effective. There are also prototype designs for zinc-air fuel cells, which can function as large, rechargeable batteries, and are as efficient as any battery based system, but with ranges around 400 to 500 miles. For long trips, the electrolyte can even be completely replaced/exchanged at filling stations, which then recharge and recycle the spent electrolyte.

Infrastructure

Since hydrogen causes hydrogen embrittlement of steel, it is not clear if hydrogen can simply be put into today's natural gas transmission systems. Proponents of the hydrogen economy envision local hydrogen sources. The challenges that large, rural high-efficiency hydrogen generators face are far more acute in an urban environment. Thus, some kind of transmission system will probably be required for cities.

Hydrogen use would require the alteration of industry and transport on a scale never seen before in history. For example, the distribution of hydrogen fuel for vehicles in the U.S. would require an entirely new infrastructure costing hundreds of billions of dollars, or more. However, it is believed that future oil costs, poor alternatives and improvements in technology may make the transition economically viable in the future.

Cost

Hydrogen seems unlikely to be the cheapest carrier of energy over long distances in the near future. Advances in electrolysis and fuel cell technology have not addressed the underlying cost problem yet.

Hydrogen pipelines are more expensive than even long-distance electric lines. Hydrogen is about three times bulkier in volume than natural gas for the same energy delivered, and hydrogen accelerates the cracking of steel (hydrogen embrittlement), which increases maintenance costs, leakage rates, and material costs. The difference in cost is likely to expand with newer technology: wires suspended in air can utilize higher voltage with only marginally increased material costs, but higher pressure pipes require proportionally more material.

Setting up a hydrogen economy would require huge investments in the infrastructure to store and distribute hydrogen to vehicles, in addition to the cost of the new vehicles themselves. In contrast, battery electric vehicles, which are currently at a similar stage of technological maturity, don't require infrastructure investments. Since hydrogen is likely to be produced with the same sources as electricity (fossil, nuclear, solar, wind) it may be less economical than a pure electricity economy.

How would Hydrogen be produced? Energy would/could come from a multiple of sources i.e. Natural gas, nuclear, solar, wind, biomass, coal, other fossil fuels, and geothermal.

Natural Gas

Uses steam reformation. Requires 15.9 million cubic feed of gas—that's 777,000 facilities and would cost 1 trillion dollars to cover 150 million tons of Hydrogen gas annually, the estimate for consumption in 2040. $ 3.00 per GGE (Gallons of Gas Equivalent).

Nuclear

Provides energy for electrolysis of water. Would require 240,000 tons of un-enriched uranium—that's 2000 600-megawatt power-plants, which would cost $ 840 billion about $ 2.50 per GGE.

Solar

Provides energy for electrolysis of water. Would require 2500 kilowatt-hours of sun per square meter 113 million 40 kilowatt systems, which would cost $ 22 trillion about $ 9.50 per GGE.

Wind

Provides energy for electrolysis of water. At 7 meters per second average windspeed, it would require 1 million 2 MW wind turbines, which would cost $ 3 trillion about $ 3.00 per GGE.

Biomass

Gasification plants would produce gas with steam reformation. 1.5 billion tons of dry biomass 3300 plants which would require 113.4 million acres of farm to produce the biomass. $ 565 Billion in cast or about $ 1.90 per GGE.

Coal

FutureGen plants use coal gasification then steam reformation. Requires one billion tons of coal or about 1000 275-megawatt plants cost of about $ 500 billion about $ 1 per GGE.

Alternatives to the Hydrogen Economy

Hydrogen is simply a method to store and transmit energy. Various alternative energy transmission and storage scenarios may be more economic, in both near and far term. These include:

Compressed air

Solving many of the generation, transportation and storage problems which plague hydrogen, compressed air suffers from a low energy density. Because compressed air is the direct storage of mechanical energy however, comparatively little energy is lost in conversion during both generation and use.

The electrical grid plus batteries

The electrical grid and chemical storage battery pose viable long-term alternatives to hydrogen in transmission, especially as more large batteries are made available on the grid in the form of electric or hybrid autos, which might act as load-balancers. The solar cell might also be used in some areas to make energy locally for battery powered autos. Of these technologies, only grid power is currently in a high state of technical development. Solar power suffers from a low power density to area, making it difficult to use in transport. High capacity batteries (chemical cells) have already seen use in commercial hybrid cars, but these have yet to be used in load-balancing. It is possible that a combination of battery and hydrogen power will be used in the future, although many think that hybrid cars running on battery power and green fuels is a more viable option.

Hydrogen production of greenhouse-neutral alcohol

Hydrogen in a full "hydrogen economy" has been envisioned as a way to make renewable energy available to automobiles which are not all-electric. A final theoretical alternative to hydrogen would do this by using hydrogen locally (captive use) to make liquid fuels from a CO_2 source. To be greenhouse-neutral, this source would be from air, biomass, or from CO_2 which would otherwise be scheduled to be released into the air from non-carbon-capture fuel-burning powerplants (of which there should still be many in the future, since economic carbon capture and storage is site-dependent and difficult to retrofit). These alcohols would then act as greenhouse-neutral *additional* energy stores and carriers for transportation, but without disrupting present methods of liquid fuel transport and use. Rather than be transported from its production site, hydrogen may thus instead be used centrally/locally to produce renewable liquid fuels which may be cycled into the present transportation infrastructure directly, requiring almost no infrastructure change.

Hybrid Strategy of Electricity and Synthetic Methane

Electricity can be more efficiently used in a storage battery than electrolysing water to hydrogen. For example, a storage battery may retain about 90 per cent of the electricity used to charge it, and be able to provide about 90 per cent of the electricity that it can store, resulting in a "round trip" efficiency of about 80 per cent. This is compared with a 70 per cent efficiency of electrolyis and perhaps 60 per cent efficiency of a fuel cell, resulting in a round trip efficiency of only about 40 per cent for hydrogen—only about half the efficiency of batteries.

But batteries are expensive, and they wear out over time. The cost of energy storage in batteries is drively largely by the cost of the batteries themselves, and not the cost of the energy put into them. The cost of storing energy for any extended length of time is prohibitive when using batteries, even when compared with hydrogen storage.

On the other hand, the cost of energy storage of hydrogen could be reduced further by using the hydrogen (and carbon dioxide) to synthesize methane, using a Sabatier reactor. This process is about 80 per cent efficient, reducing the round trip efficiency to about 20 to 30 per cent, depending on the method of fuel utilization. This is even lower than hydrogen, but the storage costs drop by at least a factor of 3. Methane is 3.2 times denser energetically than hydrogen, is easier to store—and an infrastructure (Natural Gas pipelines) are already in place. The advantage of methane storage is that it is very inexpensive, once one has accepted the high cost of conversion.

We can start to see how a hybrid strategy could be more effective than hydrogen alone. Short-term energy storage (meaning the energy is used not long after it has been captured) is best accomplished with battery or even ultracapacitor storage. Longer term energy storage (meaning the energy is used weeks or months after capture) is best done with synthetic methane, which can be stored indefinitely at relatively low cost. The strategy dovetails well with the recent interest in Plug-in Hybrid Electric Vehicles, or PHEVs, which use a hybrid strategy of electrical and fuel storage for their energy needs.

Hydrogen storage is therefore only optimal in a narrow range of energy storage time, probably somewhere between a few days and a few weeks. This range is subject to further narrowing with any improvements in battery technology. It is always possible that some kind of breakthrough in hydrogen storage or generation could occur, but this is unlikely given the physical and chemical limitations of the technical choices are fairly well understood.

Summary

Hydrogen can replaced the other fuel if it is possible to overcome the present limitations. Solar photovoltaic hydrogen may solve the problem. The concept of "energy-plus home" is the real solution. Future energy can (should) include PV and hydrogen. The concept of the zero-energy building can be envisioned to expand to the "energy-plus home" that produces more energy (electricity for the residence, hydrogen for night-time power and the family "Freedom Car"). The solar-hydrogen park or village is an extension of this, in which the solar energy and hydrogen is shared in the community. Again, additional electricity can be supplied to the grid and any excess hydrogen can be sold through the community's refueling stations. The marriage between hydrogen and solar brings secure, clean energy as well as making PV a "24-hour" option.

REFERENCES

1. P. Maycock, *PV News*, Vol. 23(3), Mar. 2004; Photon International, April 2004, 40-41.
2. See, for example, technical reports in the Proceedings of this conference, (2004).
3. L.L. Kazmerski, *Proc. 39th IEEE Photovoltaics Spec. Conf.*, (IEEE, New York; 2003), 21-27.
4. T. Surek and L.L. Kazmerski, SPIE *oe Magazine,* (SPIE, Bellingham, WA), May 2003, 24-27.
5. J. Perlin, L.L. Kazmerski, and S. Moon, *Solar To-day*, Jan.-Feb. 2004, 24-27.
6. G. Davis, "Evolving Sources or Revolutionary Technology—Exploring Alternative Energy Paths to 2050", *Proc. Oil & Money Conference*, London, Oct. 2001, 1-11.
7. National Vision of America's Transition to a Hydrogen Economy to 2030 and Beyond, Feb. 2002.
8. National Hydrogen Energy Roadmap, Nov. 2002.
9. Hydrogen, Fuel Cells & Infrastructure Technologies Program, Multi-Year Research, Development and Demonstration Plan (U.S. Department of Energy, Energy Efficiency and Renewable Energy, Washington, DC), June 2003.
10. *Scientific American*, May 2004, 66-74.
11. National Energy Policy, Report of the National Energy Policy Development Group, May 2001.
12. U.S. DOE Solar Energy Technology Program Multi-Year Technical Plan, 2003-2007, (NREL, Golden, CO; 2004).
13. J. Deutch and E. Moniz, Editorial, "N-Power Key to World's Future", *Arizona Daily Star*, Friday, August 15, 2003 (www.azstarnet.com).
14. Energy Information Administration, International Energy Annual Report, 2002, (Washington, DC).
15. L.L. Kazmerski, *Solar Today*, Vol. 16, No. 4, 2002, 40-43.
16. L.L. Kazmerski, *Proc. Hybrid Conf.* (2004). Also, PV FAQ (www.nrel.gov/ncpv).
17. PV FAQ sheets, www.nrel.gov/ncpv.
18. NAC, *The Hydrogen Economy: Opportunities, Costs, Barriers, and R&D*, (National Academies Press, 2004).
19. Solar Electric Power: U.S. Photovoltaics Industry Roadmap, (U.S. PV Industry; May 2001).
20. J.A. Turner, *Science*, Vol. 285, July 30, 1999, 687-89.

21. See, for example, *New York Times*, August 15, 2003.
22. M.K. Mann, P.L. Spath, and A.S. Watt, 9th Canadian Hydrogen Association Meeting, Vancouver, B.C. (Hydrogen Association).
23. R. Rocheleau, E. Miller, and A. Misra, Proc. U.S. DOE Hydrogen Prog. Rev. NREL/CP-430-21968, (National Renewable Energy Laboratory, Golden, CO; 1996), 345-52.
24. J.M. Vidueria, A. Contreras, and T.N. Vesiroglu, *Int. J. Hydrogen Energy*, Vol. 28, (2003), 927–37.
25. D.L. Block and I. Melody, *Int. J. Hydrogen Energy*, Vol. 17, (1992), 857.
26. A. Kazim and T.N. Veziroglu, *Renewable Energy*, Vol. 24, (2001), 259-74.
27. C.C. Elam, C.E. Gregoire Padro, G. Sandrock, A. Luzzi, P. Lindblad, and E. Fjermestad Hagen, *Int. J. Hydrogen Energy*, Vol. 28, (2003), 601-07.
28. T. Surek, *Proc. 3rd World Conference* on Photovoltaic Energy Conversion, Osaka, Japan, May 2003.
29. R. McConnell, S. Moon, and L.L. Kazmerski, *Solar Today*, to be published, June 2004.
30. W.R. McKee, D.R. Carson, and J.D. Levine, *Proc. 16th IEEE Photovoltaics Spec. Conf.*, (IEEE, New York; 1982), 257-61.
31. N. Lewis, see Web page for detailed presentation http://www.its.caltech.edu/~mmrc/nsl/default.htm
32. J.J. Romm, *The Hype About Hydrogen,* (Island Press, New York; 2004).

Ecological Sanitation

3

Mohammod Aktarul Islam Chowdhury

Introduction

Either pit latrine or flushing toilet is commonly used as the main sanitation option throughout the world especially in developing countries now-a-days. Conventional waterborne sewerage system have proven to be inappropriate to solve sanitation needs in developing countries; the systems are too costly to be provided to all, and only wealthier upper rich and middle-class are normally provided with such services that estimated to be only about 10 per cent while approximately 90 per cent of the sewage in urban areas of the developing countries is untreated and discharged to the rivers, lakes and other water courses, low-lying areas of the vicinities and coastal areas where available polluting the surface water sources. Used pit latrines have the limitations especially in the densely populated areas, with severe risks of contaminating groundwater. Every year billions of rural people are being migrated to cities and towns of developing countries in pursuit of better living without providing additional sanitation facilities. Furthermore many of the rapidly growing cities and towns are located in arid and semi-arid areas where water scarcity is severely affecting the water supply to the increasing people of the urban centers. Faced with the reality of food insecurity, decreasing soil fertility and increasing prices of fertilizers in the world market, there is need to use the nutrients available in human urine and feces for agricultural purposes. Human urine and feces are rich in nitrogen and phosphates which may be used as valuable fertilizers to increase the productivity of food crops. In a situation of dreadful scenario but dreamed with future prospect, it is obvious that enormous new challenges leads to need to re-think, a crying need to raise the status of sanitation with new approaches, techniques and methods in order to use the nutrient value of the human urine and feces. In this regard, an alternative to conventional sanitation called 'ecological sanitation' based on an ecosystem approach should be tried to introduce to treat human excreta (both of the urine and feces) as the available resource to be recycled.

The Challenge of Sanitation Practices in Third World

In the developing countries of third world, the sanitation practices that are promoted today fall into one of two broad types:

(a) **Flush-and-discharge:** Over the past hundred years the flush-and-discharge has been used as the ideal technology particularly for urban areas. Many municipalities in developing countries often with the help of international financing agencies try to install this model. In a flush-and-discharge system, a relatively small amount of dangerous material such as human feces is allowed to pollute a huge amount of water. Black-water from toilet and urinal, grey-water from bathroom and kitchen, storm water from roadside and surface drain and industrial wastewater in the urban centers are flushed, collected and discharged through this system. In most cases, the resulting sewage is discharged completely untreated into surface water bodies in many of the urban areas in developing countries. Flush-and-discharge approaches can work well and achieve an acceptable level of pathogen destruction. In the third world, sewage is nearly always discharged into the environment at large without treatment (UNCSD, 1997). Sewage discharges from centralized water-borne collection systems are a major component of water pollution contributing to the nutrient overload of water bodies, toxic algal blooms such as red tides, and adversely influencing tourism in some coastal areas (Cosmer and Thornton, 1989; Platte, 1995). Although such systems are acceptable to the vast majority of people, they are not simple and require institutional capability and technical skills generally not available in many cities of developing third worlds. Most urban growth in the developing countries are being taken place in informal settlements where municipal authorities are unwilling or unable to provide adequate sanitary services such as water supply, sewerage system, urban drainage and collection of garbage and refuge. Effective sewage treatment is so expensive that it is rarely achieved in practice, particularly in the fast growing urban centers of developing countries. Consequently, low-income households cannot rely on flush-and-discharge system; rather depend on some kind of drop-and-store sanitation technology to deal with their needs.

(b) **Drop-and-store:** For those without access to flush-and-discharge, the conventional alternative is a drop-and-store device, usually a pit latrine, based on containment and indefinite storage of human excreta. Drop-and-store is often regarded as interior temporary solution compared with flush-and-discharge especially in rural and poor peri-urban areas. Drop-and-store requires access to the ground, a reasonable amount of open space, soil that can be dug, a low groundwater level and a site that is never flooded. Drop-and-store systems can be simple without the requirement of water for flushing and relatively low-cost using any material such as paper, solid object or water for anal cleaning. But they have many drawbacks *vis-à-vis* soil and groundwater contamination, bad odor, fly breeding, collapse of pit, possibility of destabilization of nearby buildings, risk for overflow during heavy rains. Often drop-and-store systems cannot be used at all in crowded areas, on rocky ground where the groundwater level is high and in areas periodically flooded. However, drop-and-store technologies can prevent pollution in some places, but they are often not feasible in urban areas because of lack of space for digging deep pits, difficult soil and groundwater conditions, destabilization of foundations of nearby houses.

Thus leaders and communities are currently faced with many limitations and weakness in both of the two above mentioned options of sanitation systems and seek entirely new solutions because existing approaches are not viable and affordable to the vast majority of the people, neither do they offer people a sustainable approach. Ecological sanitation with a new vision and criteria may an alternative towards sustainable society.

Ecological Sanitation (Eco-san), a New Approach

The new approach to sanitation called ecological sanitation or 'eco-san' is a cycle a sustainable closed loop system based on three fundamental aspects:

- Rendering human excreta safe
- Preventing pollution rather than attempting to control it after we pollute it, and
- Using the safe products of sanitized human excreta for agricultural purpose, plantation, eco-culture and aqua-culture for living bodies other than human.

Ecological sanitation replaces nature by returning sanitized human urine and feces to the soil instead of polluting the environment; human urine and feces are used to improve soil structure and supply nutrients.

Eco-san is an alternative approach to conventional sanitation aiming to a sustainable system in the form of a cycle that helps to protect water resources and enriches the soil. It is an ecosystem approach that regards human urine and feces as valuable resources to be recycled (Forum, 1999).

Eco-san represents a vision of sustainable sanitation systems which are based on a systematic material-flow-oriented recycling process that constitutes promising, up-to-date, holistic alternatives to conventional solutions. Ideally eco-san system permit the complete recovery of all nutrients from feces, urine and grey-water, benefiting agriculture and minimizing water pollution, as well as allowing economic use of water and its maximal reuse particularly for irrigation purpose (Werner, 2001).

Key feature of ecological sanitation are prevention of pollution and disease caused by human excreta, treatment of human excreta as a resource other than waste, and recovery and recycling of the nutrients especially valuable for plants and essential for building healthy soil. On the other hand, conventional sanitation approaches misplaces such nutrients, disposes of them and break this cycle.

The Criteria of Ecological Sanitation

An ideal system of sanitation preferably ecological sanitation that contributes toward the goal of equity in society and a sustainable society must meet or on the way towards meeting the following criteria:

- Prevention of disease
- Affordability
- Protection of environment
- Acceptability
- Simplicity.

The Main Component of Ecological Sanitation

The main components of ecological sanitation are:

- Nature including climate (humidity, temperature), water (quantity, groundwater level) and soil (stability, permeability, pickability)
- Society includes settlement pattern (concentration/ dispersed, low/high rise), attitudes (fecophobic/ fecophilic), habits (washers/wipers), beliefs and taboos related to human excreta as well as economic status of the community
- Process may be of physical, biological and chemical
- Device mean the on-site structures specially built for defecation and urination.

Ecological Sanitation—An Ecosystem Approach

Ecological sanitation presents a shift in the way people think about and act upon human excreta. It is a system intelligently designed and constructed to mimic nature at every step. Eco-san is an ecosystem approach, a move away from a linear to a circular flow of nutrients—towards a recycling society. In eco-san, nutrients and organic matter in human excreta are considered a resource, food for a healthy ecology of beneficial soil organisms that eventually produce food or other benefits for people. This new approach recognizes the need and benefit of protecting environmental health and promoting human well-being, recovering and recycling nutrients, and conserving and protecting natural resources. It represents a closed-loop approach to nutrients and water problems defining a safe, green and valuable cyclic process as shown in Figure 1 (Esray, 2001).

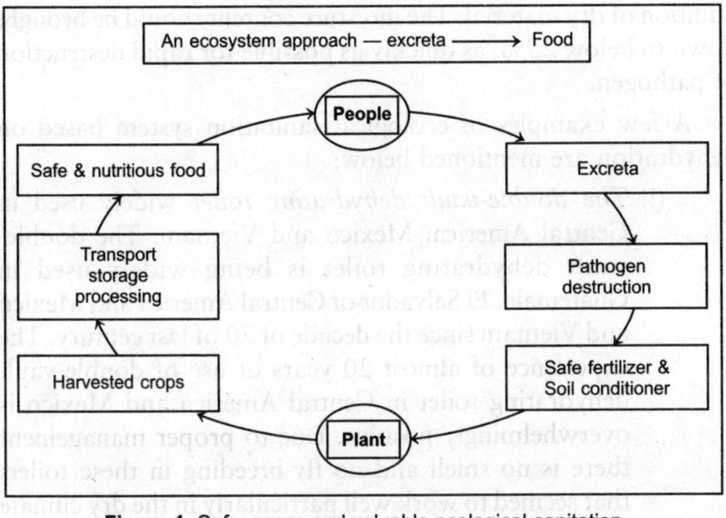

Figure 1: Safe, green and valuable ecological sanitation

Eco-san prevents pollution, destroy pathogenic organisms and recycle human excreta as fertilizer. Thus it is an ecological subsystem—part of a larger ecological system, part of the biosphere which has the components relating to nature, climate, soil, water and vegetation related to culture, beliefs, and behavior, municipal and household economy containing technical components: squatting pans, pipes, fans, solar heater, containers, vehicles etc. (Winblad, 2001).

Ecological Sanitation System Practice

The variety of ecological sanitation system is available around the world from ancient time to today. The ecological sanitation system makes it possible, in most cases, to find one that is culturally acceptable with regard to relative affordability, expenses possibly low cost, sophistication, balance of trade-off between cost and operation. In ecological sanitation system practice, the main processes used to achieve pathogen are either dehydration or decomposition.

(a) *Ecological sanitation system based on dehydration*

In a dehydrating toilet, the contents of the processing units such as vault are dried with the help of heat, ventilation and the addition of dry material. The moisture content should be brought down to below 25%, as quickly as possible for rapid destruction of pathogen.

A few examples of ecological sanitation system based on dehydration are mentioned below:

(i) *The double-vault dehydrating toilet* widely used in Central America, Mexico and Vietnam: The double-vault dehydrating toilet is being widely used in Guatemala, El Salvador of Central America and Mexico and Vietnam since the decade of 70 of last century. The experience of almost 20 years of use of double-vault dehydrating toilet in Central America and Mexico is overwhelmingly positive. Due to proper management there is no smell and no fly-breeding in these toilets that seemed to work well particularly in the dry climate of Mexican highlands. Where the system has failed with regard to wetness in the processing chamber, odor, fly breeding etc., it was usually due to non-existent, weak or bungled information, training and follow-up. In El Salvador, the system has been successfully used in a poor, high density squatter area; and by a number of middle class families living in modern, high standard houses of urban areas. The careful management of ecological sanitation system, resulting from high motivation and understanding on the part of the families involved, can make an extremely simple technology work very well. In Vietnam, the experience is mixture; when properly used, does function very well. A retention time of 2-month seems to be short for total pathogen destruction. The real problem is that some farmers tend to empty the processing chambers whenever they need the fertilizer, regardless of the retention time meaning that partly processed and even fresh feces are spread on the fields taking a long time to break age-old habits. Obviously there is a definitely a need to intensify and continue hygiene education and the practice of using fresh feces as fertilizer that can be halted.

(ii) *The 'WM Ekologen' dehydrating toilet* used in Sweden: The 'WM Ekologen' type ES system dehydrating toilet developed by Karolinska Institute, Stockholm in the early 80's of last century was widely used in Sweden. The *'WM Ekologen'* is a well tested high standard sanitation system suitable for indoor use in modern bathroom. It is used in urban as well as rural areas and in government institutions, offices as well as in private households.

(iii) *The Tecpan solar heated toilet*: The Tecpan sanitation system is developed based on dehydration, urine diversion and solar heating in a single vault device. The Tecpan solar heated toilet has been widely used in El Salvador, Mexico, and Tanzania since the decade of 90 of last century and more recently in Vietnam. The Tecpan toilets functions very well. The solar heater to the processing chamber accelerates the dehydration process and destruction of pathogen. Most of the toilets are completely dry and odor free and there is no fly breeding.

(iv) *The double-vault solar heated toilet*: The double-vault toilets with solar heated lids have been used in Ecuador since 1985. In this type of toilets a recycling system was chosen in an attempt to help the chronic problems of falling soil fertility in the high altitude region ranging from 3500 to 4000 m. The experience of the double-vault solar heated toilet in Ecuador is interesting in that it shows that the extremely dry climate of the Andean mountain creates condition which eliminates the need for urine diversion as well as solar heat collectors. In humid climate, urine diversion as well as solar heat collectors were added in an attempt to speed up the evaporation from the processing vault.

(v) *The indoor dehydrating toilet:* The indoor dehydrating toilet is used in Ladakh of India, a dry highland region in the western Himalayas at an altitude of 3500 m. In this type of toilet, it is possible to dehydrate the feces without prior diversion of urine due to the dry climate. As long as the toilet is well-maintained and enough soil is pushed down the drop hole every day, there are no odor; in some cases there is a faint smell of ammonia from urine splashed on the soil-covered floor of the toilet room. There is no fly-breeding due to the dryness of the soil/excreta pile. The system has worked well in rural areas for hundreds of years but in recent years there have been problems where households have no easy access to soil.

(vi) *The indoor long-drop dehydrating toilet*: The indoor long-drop dehydrating toilet system is a traditional system has been widely used in Yemen for hundreds of years. This is an example of the dry eco-sanitation approach applied to an urban situation with multi-storey housing and communal collection of dehydrated feces by special staff in a culture where people are washers. This is free of smell and fly-breeding where urine and cleaning water are evaporated easily. In indoor long-drop dehydrating toilet feces are processed in three steps: first they are dehydrated on site, second they are further dehydrated and subject to direct solar radiation at the public bath, and finally they are burnt.

(b) *Sanitation system based on decomposition (or composting)*

Composting is a biological process in which a variety of micro-organisms break down organic matters to humus under controlled conditions. Microorganisms such as viruses, bacteria, fungi, algae to worms and insects take part in the breakdown of organic materials in a composting toilet. Humus is a rich stable medium in roots thrive. In a composting toilet, human excreta along with additional bulking agents are deposited into a processing chamber where soil-based microorganisms decompose the solids just as eventually happens to all organic matters in the environment. Additional bulking agent includes vegetable scraps, straw, peat moss, wood shavings, coconut husks etc. Temperature, airflow and other factors are controlled to varying degrees to promote optimal conditions for composting. The humus produced in the composting toilet is an excellent soil-conditioner, free of human pathogens when the proper conditions are achieved adequate retention time is allowed. Odor, if any, can be extracted directly out above the roof of the toilet through a ventilation system.

In a composting toilet, the following optimal conditions for biological decomposition should be achieved:

- The sufficient oxygen should be able to penetrate the compost heap to maintain aerobic condition;
- The material in the composting unit or vault should have a moisture content of 50 to 60 per cent;
- The carbon-nitrogen ratio should be within the range of 15:1 to 30:1; and
- The temperature of the composting vault should be above 15°C.

A few examples of ecological sanitation system based on decomposition or composting are mentioned below:

(i) *The 'Clivus Multrum' single-vault composting toilet:* The 'Clivus Multrum' single-vault composting toilet was introduced in Sweden for use in weekend houses during 1950s. Since then a wide variety of models have come to the market and are being used in different parts of the world including Australia and North America. Now-a-days Clivus Multrum is not only used in weekend houses but also in regular houses, in institutions and as public toilets. About 10,000 Clivus Multrum toilets are in use worldwide.

The maximum number of users of Clivus Multrum depends on a few factors such as temperature, humidity, amounts and type of refuse, proportion of urine to feces, volume of the receptacle etc.; in most cases, the maxim number of users for one Clivus Multrum in regular year-round use is 8 to 10 people. The humus from the Clivus Multrum has a similar bacterial content to soil and it is considered safe to use directly as a fertilizer and soil conditioner for the cultivation. If the Clivus Multrum is properly built, then it is reasonably free of nuisance.

(ii) *The 'Carousel' multiple vault composting toilets:* The 'Carousel' multiple vault composting toilet first manufactured by Vera Miljo A/S of Norway has been used in Norway and USA since 1972. A similar type of toilet is also manufactured and used in Sweden. Outside Scandinavia another Carousel type of toilet named 'Rota-Loo' is manufactured and used in Australia, New Zealand and some islands of South Pacific. The carousel is basically a multi-vault toilet. It effectively keeps fresh and sanitized feces separate.

(iii) *Sirdo Seco solar heated composting toilet:* Sirdo Seco solar heated composting toilet is a prototype double-vault composting toilet normally heated by sun light the main source of solar energy. Sirdo Seco solar heated composting toilet was first tested in Tanzania in the mid-1970s further developed in Mexico and Vietnam. In Mexico it has been successfully used since 1980. When properly managed, the toilet works well with a high capacity. The toilet is low cost too being the cost of a fabricated toilet around US $100. One particular advantage of this lightweight toilet is that it is mobile and can be easily made with fiberglass or thick plastic sheet. People living in a squatter slum when needed to be evicted at a short notice, they can arrange to have the toilet emptied and take it with them like a piece of furniture. This toilet is suitable for a family of 6 to 8 people which need to empty once a year. Such type of toilet can be easily used in the wetlands like the haor areas of Bangladesh.

(iv) *The movable bin toilet:* The Movable Bin Toilet was developed and tested by the Centre for Environmental Studies at the University of Tasmania of Australia. Incorporation with the local counterpart, they have successfully tested several movable composting toilets in the Pacific Island of Kiribati. This toilet became very successful in producing an innocuous humus-like residue. The toilet is too cheap cost not being greater than US $50. One main advantage of this toilet is that its two composting chambers used alternately are made of thick plastic sheets or plastic refuse bins and can be moved from one place to another easily. Such type of toilet is very suitable for wetlands such as the haor areas of Bangladesh.

(v) *The CCD toilet in South Pacific:* The CCD toilet was developed by Greenpeace and Centre for Clean Development in the small islands states of South Pacific including Micronesia. It focused the sanitation system based on zero-discharge rather than diversion and recycling of urine. The CCD toilet consists of two water-tight chambers used as composting vaults built above ground. The experience indicates that with a system of CCD type toilet it is possible to a degree of liquid evaporation and maintenance free operation generally not available in composting toilets in a humid environment. The CCD toilet is promising as an appropriate sanitation solution where environmental contamination is a major concern, and even in cultural settings in which a high level of maintenance is not likely to be expected, provided there is a supply of appropriate organic bulking agents such as leaves, vegetable scraps, coconut husks or wood shavings.

(vi) *Double-vault toilet connected to the reed-bed:* Double-vault toilet connected to the reed-bed has been adapted to the population of 'washers' in Kerala of India. Not only urine, but also the water for anal-cleansing is diverted into an evapo-transpiration reed-bed next to the toilet. The vault is lined with straw before use which provides a carbon-rich bed to receive the feces and also to absorb moisture. A handful of ashes are sprinkled over the feces after each use. Occasionally some straw, leafy material and paper scraps are also added ensuring a process of decomposition rather than dehydration. Such sanitation system became successful due to good mobilization of the local population, specially the women, to effective hygiene education and to regular follow-up.

IMPORTANT ASPECTS OF ECOLOGICAL SANITATION

Unfamiliar Aspects

Probably the most unfamiliar aspect of ecological sanitation is that it requires *some handling of the products* at the household level. A most important point about handling is that once ecological sanitation has gone to scale and hundreds or thousands of units are in use in towns and cities, individual households no longer need to handle the products at all.

A second cultural issue, which causes debate, is whether eco-san toilets will be used properly in cultures where washing after defecation is mandatory by tradition and religion.

Often people are unfamiliar that eco-san system need not be expensive to build because:

- The entire sanitation device is built above ground—there is thus no need for expensive digging and lining of pits
- Urine is diverted, no water is used for flushing and the volume of the processing chamber(s)/vault(s) is fairly small
- the contents of the processing chamber(s)/vault(s) are dry that means there is no need for expensive water-tight construction.

Also people unfamiliar with eco-san options cannot imagine how they can be used in multistory buildings.

Sensitivity of Eco-san Toilet

Eco-san toilets are more complex than drop-and-store system but infinity less complex than flush-and-discharge system. In spite of the potential advantages, there is the risk that those who plan, design and build always do not fully understand the basic principles involved and how they relate to the local conditions. Even with the right system in place, the most common reasons for failure of the eco-san system are:

- Lack of participation from the users
- Want of understanding of how the system works
- Defective materials
- Poor workmanship and
- Improper maintenance.

Design and Management Features of Eco-san

The most common features of eco-san regarding design and management are:

- Dealing with liquids
- Urine diversion
- Liquid separation and
- Combined processing.

Handling of Grey-water or Sullage

The water generated from food preparation, bathing and washing is known as grey-water or sullage. Dry sanitation system, normally practiced for eco-san, does not handle grey-water and therefore a separate system must be devised to take care of the grey-water.

The amount of grey-water can be significantly reduced through behavioral changes, good maintenance of pipe and water taps and the use of water-saving devices.

Choosing an Eco-san System

Ideally an eco-san system will prevent pollution, sanitize external nutrients and return them to the soil like the arrangement 'sanitize and recycle' and require no water for transport or processing. Eco-san system is normally chosen to address the issues of greatest concern in the place where they are used. A number of local variables cited below influence the choice of an appropriate eco-san system:

- Climate
- Topography and type of soil
- Abundance/scarcity of water
- Proximity/sensitivity of water resources and aquatic ecosystem
- Energy
- Social and cultural factors
- Economy
- Technical capacity
- Infrastructure
- Population density and settlement pattern
- Agriculture etc.

Promotion and Support of Eco-san to Families and Communities

Eco-san is more complex than conventional sanitation systems and usually place more responsibility for appropriate functioning on individual families and local communities. The important critical issues related to promoting and supporting eco-san are:

(i) *Community empowerment and promotion:* The important critical issues related to community empowerment and promotion are:

- Needs based approach
- Promotion strategies
- Technology selection and adaptation and
- Phasing the program.

(ii) *Education and training of eco-san facilitators and communities:* The important issues to be considered for education and training of eco-san facilitators and communities are:

- Participatory learning
- Sharing information and
- Skills training for key local authorities, field staff and field workers as well as household and community members.

(iii) *Institutional framework:* Various types of institutions and organizations need to be involved in promoting and supporting eco-san system. Following organizations, bodies, sectors and institutions may be effectively involved in this regard:

- Local and non-governmental organizations (NGOs)
- Government and official bodies
- Commercial and private sectors
- International development organizations and donors
- Educational and research institutions
- Technical schools and technology centers
- Professional associations and bodies.

Integrated strategies and partnerships among the involved organizations and institutions have to be developed for the strong eco-san programs.

(iv) *Financial Considerations:* The introduction of eco-san system is bound to lower the costs of traditional and conventional sanitation system whatever be the urban or rural. Eco-san systems will involve the costs for:

- Information
- Training
- Monitoring
- Follow-up
- Safe-handling of urine
- Transportation of urine
- Storage of urine
- Dehydrated or composted materials etc.

The economic value of the fertilizers or compost produced from successful eco-san system should also be considered to ensure sound finance. Households have to repay not only full operation and maintenance costs but also to be ready to refund the investment cost as much as possible. However, subsidies may be needed to attract the communities at the initial stage of introducing the system in a locality.

Advantages of Ecological Sanitation

The main advantages of eco-san are:

(i) *Advantages to the environment:* If eco-san could be adopted on a large scale, the following advantages would be added to the environment:

- Protection of water sources including groundwater, streams, lakes, rivers and the sea from fecal contamination
- Less water consumption
- Requirement of less amount of expensive chemical fertilizers by the farmers
- Less environmental degradation to soil and water
- Possibility of making use of the high fertilizer value of urine by dilution of urine with water followed by direct use on vegetable gardens and agricultural fields
- Recovery of the resource value of feces by turning into valuable soil conditioner destroying the dangerous pathogen micro-organisms
- Reduction of green-house effect by large scale recycling through the increase of carbon content of soil
- Rejuvenating rural and urban agriculture by returning human urine and sanitized feces to replenish the soil nutrient.

(ii) *Advantages to households and neighborhoods:* If eco-san managed properly, the following advantages would be contributed to households and neighborhoods:

- No smell, no fly nuisance and no insect breeding due to no scope of coming into contact of urine and feces into open atmosphere
- No risk of falling of small children into the eco-san toilet because they are neither deep nor wet and usually built entirely above ground

- Scope of improving unpleasant environment immediately by the individual household
- Privacy, aesthetic advantage and the convenience of the households can be ensured through eco-san
- Decrease of likelihood of diarrhea and other worm infections leading to better health
- Improving the nutrition condition of the family if the urine and feces are recycled to grow vegetables in garden plots, on rooftops and balconies
- Being lightweight and movable, eco-san toilets can be shifted from one place to another if needed
- Eco-san based on dehydration or decomposition by composting reduces the volume of material to be handled and transported and result in a dry, soil-like, completely inoffensive and easy-to-handle product
- Eco-san neither destabilizes the foundation of nearby house or building nor collapses.

(iii) *Advantages to municipalities:* If eco-san maintained properly, the following advantages would be shared to municipalities:

- Eco-san does not allow the use of scarce water resources and may result in a more equitable allocation of water to both of the rich and poor households
- Eco-san has the potential to increase sanitation coverage of the unserved people more quickly than any other method
- Eco-san is generally affordable to the poor
- Eco-san has almost no recurrent costs for operation and maintenance
- Eco-san requires no excavation
- Eco-san does not depend on water and pipe network
- Eco-san can be placed anywhere even inside a house and on upper floors
- Eco-san can be used even in congested areas

- Eco-san could be an expensive and an alternative to the expansion of sewerage system
- Eco-san can allow, even favor decentralized urban waste-to-resource management.

Conclusion

In order to improve the health of the world's population increasingly threatened by inadequate and unhygienic sanitation as well a number of forthcoming crisis, ecological may an effective solution by reducing the risk of pathogens being spread from one area to another, avoiding mixing urine and feces, and avoiding mixing excreta and water preventing pollution as well as providing the benefits such as conserving and protecting water resources, sustainable food production by the appropriate use of human and animal urine and feces sanitation, contributing to the reduction of greenhouse effects. Obviously eco-san should be improved through further research, adopted and adapted throughout the world especially in the areas of high population density like South Asian Sub-continant including Bangladesh and India.

REFERENCES

Cosmer, P. and Thornton, J. (1989). "We All Live Downstream—The Mississippi River and the National Toxic Crisis", *Greenpeace*, Washington DC, USA.

Esray, S.A. (2001). "Towards a Recycling Society Ecological Sanitation—Closing the Loop to Food Security" in the proceedings of the International Symposium on 'Ecosan-closing the Loop in Wastewater Management and Sanitation' held on October 30-31, 2000 published by Deutsche Gesellschaft fur Technische Zusammenarbelt (GTZ) GmbH, Germany, 34-44.

Forum (1999). "Ecological Sanitation—An Alternative Approach", *Water & Sanitation FORUM Newsletter* published by Philippine Centre for Water and Sanitation—International Training Network (ITN) Foundation, Vol. 4, No. 2, 12-13.

Platte, A.E. (1995). "Dying Seas", *World Watch*, 8(1), Worldwatch Institute, Washington DC, USA.

UNCSD (1977). "Comprehensive Assessment of the Freshwater Resources of the World" *Economic and Social Council*, Fifth Session, 5-25 April, E/CN, 171997/9, New York, USA.

Werner, C. (2001). "Ecosan—A Holistic Approach to Material-flow-Management in Sanitation" in the proceedings of the International Symposium on 'Ecosan-closing the loop in Wastewater Management and Sanitation' held on October 30-31, 2000 published by Deutsche Gesellschaft fur Technische Zusammenarbelt (GTZ) GmbH, Germany, 29-33.

Winblad, U. (2001). "Development of Eco-san System" in the proceedings of the International Symposium on 'Ecosan-closing the Loop in Wastewater Management and Sanitation' held on October 30-31, 2000 published by Deutsche Gesellschaft fur Technische Zusammenarbelt (GTZ) GmbH, Germany, 58-62.

Stickers, Tags, Marks, Logos, Labels? Trading Green Products

4

Simi T.B.

1. Introduction

The desire for "environment-friendly" products developed in the late 70's due to the rise in awareness for a safe and clean environment. Thence, there is a growing desire to protect the environment from adverse affects caused due to production, use, consumption, and disposal of products. This desire gradually opened the doors in the market for green, environment-friendly products, which subsequently coined out a new concept called "eco-labels".

1.1. *Origin*

Eco-labelling was first initiated by Germany in 1978 with the release of the "Blue Angel" program. They intended to enlighten consumer about the environment-friendly nature of a particular product. It evaluates, authenticate, and standardize "green" claims and inform consumers, by providing a product's overall environmental character thus conveying that the labelled product is more environment-friendly than most in its product category. It not only inform about quality of a particular product quality but also about the whole life cycle, including generation of inputs, production processes, consumption and waste disposal.

Agenda 21[1] recommend governments to promote environmental labelling to facilitate change in consumption patterns and thereby safeguard the environment for sustainable development. It explicitly mentions environmental labelling as an instrument for improving the environmental quality of products.

By the late 1980s and early 1990s over fifteen independent national and multi-national eco-labelling programs were established. Now, most of the countries, both developed and developing countries have established eco-labelling program in many different forms at local, national, regional and international level. The relevance of this subject significantly developed as a result of its implications with trade relations, economic development and the environment.

1.2. *Definition*

In its 1991 study, Environmental Labelling in OECD Countries, the Organization for Economic Co-operation and Development defined 'environmental labelling' as the 'voluntary granting of labels by a private or public body in order to inform consumers and thereby promote consumer products which are determined to be environmentally more friendly than other functionally and competitively similar products'. In other terms, a recognisable symbol granted by an independent organization and placed on certified products.

Similar such definitions were brought out by various other bodies like, to mention a few, the Secretariat of United Nations Conference on Trade and Development (UNCTAD) in 1993[2] and by GATT Secretariat in 1992.[3] To be brief, it is an attempt to use market forces to reward responsible environmental behaviour.

1.3. *Objective*

- Protect the environment and at the same encourage sustainable management of resources.
- Create awareness among consumers regarding the environmental effects of a product that is to be consumed.
- Offer incentive to manufacturers and importers for reducing adverse environmental impact of products.
- Enhance environmental standards of the commodity and at the same time encourage environmentally sound innovation.

- Give such manufacturers a competitive advantage over other manufacturers.

To be brief, eco-labelling operates as an imperative and genuine means to protect and conserve the environment, simultaneously ensuring to promote sustainable production and consumption patterns.

1.4. *Types*

Generally eco-labelling programs is of three different types:

- First party labelling, where the company by itself label their products eco-friendly (self-declaring) based on the few criteria's fixed by themselves, related to specific environmental issues.
- Second party labelling, where the Industry associations set up a certification criteria for their member products.
- Third party labelling, where an initiator (public or private) independent from the producers, distributors and sellers of the labelled products set up the criteria for eco-labelling a product.

Eco-labelling schemes of third category are preferred for the reason that the certification procedures are carried out by an impartial third party who has no commercial interest in the same. However, there are other ways through which eco-labels are classified. Like, voluntary and mandatory labels; self-declared and third party labels; single issue and life cycle based labels; single sector and multiple sector labels etc. Therefore, there are different types of ecolabels and programs.

After the initial success of the ISO 9000 standards on total quality management, the International Organization for Standardization (ISO) began its work on a new series of standards related to environmental management systems, an element of which was related to eco-labelling. Within a short span the organization came up with certain eco-standards popularly known as the ISO 14020 series. As per the series there are three types of eco-labelling:[4]

- ISO 14024 (Type I)—a voluntary, multiple-criteria based, third party program that awards a license which authorises the use of environmental labels on products indicating overall environmental preferability of a product within a product category based on life cycle considerations.

- ISO 14021 (Type II)—informative environmental self-declaration claims.
- ISO 14025 (Type III)—voluntary programs that provide quantified environmental data of a product, under pre-set categories of parameters set by a qualified third party and based on life cycle assessment, and verified by that or another qualified third party.

This series of standards, ISO 14020, have been recognised as international best practice and are an accepted approach to delivering environmental labelling and declarations on products and services.

2. Eco-labelling and Trade

Though the use of the eco-labels by governments, industry and non-governmental organizations kept mounting, there was a growing distress within the trade fraternity at both national and international level. This distress was primarily due to fact that the domestic producers would now better positioned to influence the choice of the product groups and the criteria used for awarding the labels, than the exporters. Thus holding a concern that eco-labelling would adversely affect export competitiveness and henceforth act as a non-tariff trade barrier.

Besides, these standards are seldom established in consistency with the established international standardization rules that are in a non-discriminatory, transparent and open manner. In most occasions, the foreign producers are compelled to meet criteria's that are not relevant in their country of production. For instance, the SO_2 emission during a production process. In a country where such emissions are not a matter of concern the eco-labelling requirement would impose extra costs. The foreign producers would be forced to import the technologies that are developed to deal with pollutants that are important in those countries but less important in the country of production.[5]

Hence developing countries are more vulnerable to this discriminatory trade impact, particularly in the case of small and medium enterprises in export market as most of the new product categories that are being selected for inclusion in the eco-labelling schemes are of particular interest to them. In addition the international standards are always too high for such small countries to adhere as they lack resources to undertake costly testing, verification, plant inspection, and certification procedures required for compliance with the requirements.

Other than high cost of compliance, lack of adequate knowledge about such programs also contributes itself to adverse effects. Producers in many countries have little or no information about certain ecolabel programs; they have no clue about the certification needs or where and by whom to be certified. In addition, the advanced technologies that are used to develop ecolabels are often patented. Placing the developing country producers in peril as they rarely have access to such technologies. Even if they could access such patented technologies, still the cost of such technologies are usually unaffordable.

Therefore, eco-labelling effect in both ways, it effectively address the environmental concerns with respect to production and consumption of a product but at the same time act as an effective trade barrier to all small foreign producers. In fact, the Members and Secretariats of the OECD, UNCTAD and WTO/GATT have put several proposals forward to address the trade effects of eco-labelling schemes. These include proposals for harmonization, mutual recognition and for greater transparency in the operation of labelling schemes.[6] Such a move could facilitate implementation of eco-labelling schemes in a manner that can encompass diverse levels of technological and socio-economic development and simultaneously take into consideration the exceptionality of the environmental conditions in each country, in addition to the recognition of international environmental responsibilities.

The Agenda 21 and the Rio Declaration, the two non-binding instruments, also address the environmental labelling schemes. In Agenda 21 there is a clear endorsement of environmental labelling in Chapter 4(B)(c).[7] Similar is the case

Box 1: Too Many Schemes and Parameters

One of the main problems with present eco-label schemes is their sheer number. The selected criterion for specific schemes does not have a common range and the eco-labelling schemes do not take into account particular environmental and labour conditions in individual developing countries. Additional problems for developing countries included the lack of an infrastructure for testing, auditing and verification procedures in developing countries. The high cost of the procedures, the lack of technical know-how and the difficulty of controlling the various stages of the life-cycle have also made the export efforts more difficult. The competitiveness may also be affected by higher costs incurred in compliance with the latest requirements. As regards the markets it should be remembered that at present the eco-labels are not more than a market niche and eco-labels seem so far to be primarily a commercial and marketing tool, rather than an eco-consideration. Some years ago there were many "eco-products" offered in European markets, especially in Germany where the Green movement has been stronger than anywhere else in Europe. There were "eco-jeans", T-shirts made of handpicked cotton only, bio-dynamic underwear.

In the field of eco-packaging hit may be mentioned that in the textile and clothing sector the questions related to environmentally friendly packaging are not very relevant. Nevertheless, it may be mentioned that different importing countries adopt varying eco-packaging laws and regulations so that the developing country exporters are often have to change their packaging for different markets. It is, therefore, very important that the exporters of textiles and clothing are aware of the different eco-packaging policies and their trade implications. However, the problems arising from the recycling or disposal of the packaging at the global level have so far very much concentrate in the Western world, the United States (US) market consuming almost one third of the total packaging materials in the world. The share of three EU-countries, Germany, the UK and France, is more or less at par with the US.

Source: Antero Hyvarinen, Eco-Labelling and Environmentally Friendly Products and Production Methods Affecting the International Trade in Textiles and Clothing, International Marketing Expert, Geneva, Switzerland,
http://www.intracen.org/ textilesandclothing/eco_labelling.htm

with Rio Declaration, where Principle 10 and 12[8] encompass particular importance for environmental labelling schemes. Therefore few international instruments contain vital implications for environmental labelling programs.

2.1. *Eco-labelling and WTO*

Eco-labelling was once talked over with much fanfare in the WTO within the Committee on Trade and Environment (CTE) and the Committee on Technical Barriers to Trade (CTBT). However no concrete decision could be taken other than assisting in raising the awareness of the need to make trade and environmental policies both compatible and mutually supportive.

Part of Doha Ministerial Declaration, prepared by the CTE, summarises member's opinions on this issue:

> Most Members agreed that voluntary, participatory, market-based and transparent environmental labelling schemes were potentially efficient economic instruments in order to inform consumers about environmentally friendly products. As such they could help move consumption on to a more sustainable footing. Moreover, they tended, generally, to be less trade restrictive than other instruments. [World Trade Organization, Report to the fifth Session of the WTO Ministerial Conference in Cancun, WT/CTE/8, 11 July 2003, p. 8]

Quite a few Agreements under World Trade Organization (WTO) have rules related to ecolabels. Like, the General Agreement on Tariffs and Trade 1994[9] (GATT), the General Agreement on Trade in Services[10] (GATS), the Agreement on Technical Barriers to Trade[11] (TBT), and the Agreement on Sanitary or Phytosanitary Measures[12] (SPS). The WTO issued the Code of Good Practice for the Preparation, Adoption and Application of Standards as Annex 3 to Technical Barriers to Trade Agreement in order to facilitate ecolabelling and standard setting in a manner that does not conflict with agreed international trade frameworks. Thus each of these agreements accommodates their own set of rules, some of which even overlap with rules in other agreements. However, good deal of uncertainty remains about which agreements apply to ecolabels and under what

Box 2: Viewpoints Expressed in the Committee on Trade and Environment on Eco-labelling

Most of the substantial discussions on eco-labelling in the CTE took place prior to the 1996 Singapore Ministerial Conference. This section presents the main views expressed in these discussions. (These have been extrapolated from the following summary records of CTE meetings: WT/CTE/M/5, 30 November 1995; WT/CTE/M/6, 17 January 1996; WT/CTE/M/7, 22 March 1996; WT/CTE/M/8, 11 April 1996; WT/CTE/M/10, 12 July 1996; WT/CTE/M/11, 22 August 1996; and WT/CTE/M/12, 21 October 1996.)

It is often stated that a North–South divide characterizes trade and environment discussions in the WTO, but this assertion is frequently a misrepresentation. Numerous standpoints have been taken in the CTE on the extent to which eco-labels are covered by and are consistent with WTO rules, and several proposals have been put forward on how to accommodate the trade concerns that they raise. Although it may be argued that there is a distinctly Southern perspective in the CTE on this issue, it cannot be stated that a distinctly Northern viewpoint has emerged. It is important to note that, during the CTE's discussion of this issue, a number of delegations stressed the utility of eco-labelling schemes as instruments of environmental policy.

The different positions on eco-labelling taken in the CTE have included the following:

(a) Eco-labels are both covered by and consistent with the TBT Agreement.

(b) Eco-labels are not covered by the TBT Agreement, but scope needs to be created.

(c) Eco-labels are not covered by the TBT Agreement. Creating scope for them could endanger the trading system. Tremendous care should be exercised in addressing this issue in future. A combination of increased transparency, equivalence, and mutual recognition could help alleviate their effects on trade. (Equivalence means the acceptance by a country of another country's standards or regulations as equivalent to its own, even if they are different, provided that they adequately fulfill its objectives. Mutual recognition in the context of eco-labelling schemes generally means that, if certain conditions are met, qualifying for the eco-label of an exporting country becomes an acceptable basis for the award of the eco-label used in the importing country.)

(d) Eco-labels are inconsistent with the TBT Agreement, and should not find any accommodation within the WTO system. A combination of increased transparency, equivalence, and mutual recognition could help alleviate their effects on trade.

Source: Doaa Abdel Motaal, The Agreement on Technical Barriers to Trade, The Committee on Trade and Environment, and Eco-labelling, http://www.earthscape.org/p3/sag01/P223-238.PDF.

circumstances, as these agreements were negotiated mostly without specific knowledge of or concern about ecolabelling.[13] There remain still scores of clarification on the extent to which these agreements apply, despite the fact that a number of agreements address standards and standardizing bodies.

For the WTO, the key point is that labelling requirements and practices should not discriminate either between trading partners (most favoured nation principles), or between domestically-produced goods or services and imports (national treatment).

2.2. *Eco-labelling controversy*

In 1996 the US and Canadian governments levied the argument that the voluntary labelling of products as environmentally friendly is a technical regulation that should be scrutinized under TBT. This stand was raised in the WTO Committee on Technical Barriers to Trade and the WTO Committee on Trade and the Environment.

The US position was based primarily on a proposal by a US industry coalition to severely restrict eco-labelling. However, the US environmental and consumer groups campaigned against the proposal, and the EU, Norway and Australia resisted parts of it within the WTO. The issues were never resolved; instead both the WTO committees finally adopted a neutral report.

2.3. Eco-labelling and trade in India

The mounting numbers of new eco-labelling schemes among various countries, together with the rapid changes within ecologically friendly products are generating a lot of confusion and apprehension between the producers/exporters in developing countries and transition economies.

There are concerns that the developing countries trade, especially India's, could turn out to be severely affected because of the rigid environmental measures of rich and developed nations. Particularly when the matter comes to the acquisition of international market for ecolabelled products. Since for the developing countries, as addressed earlier, it generates more costs while addressing concerns like necessary skill and technology requirement for producing such eco-friendly products. This concern is more in view of the fact that the role of the small-scale industries in export market is not so small. Nearly 60 per cent of production is by such enterprises in India therefore the contribution made by such enterprises in export is vital.

Such concerns were raised largely when such schemes affected textile and leather sectors in India, where the cost of compliance were exorbitant, mixed with difficulties in accessing technologies and so on. For instance, cost of compliance with eco-labelling schemes by Indian footwear exporters was around 33 per cent of the export price.[14] According to Piotrowski Ralph and Stefan Kratz (1999),[15] in a study carried out in India that for some firms the costs of testing for compliance with the Netherlands eco-label requirements for footwear could lead to a cost increase of up to 50 per cent.

Similar is the case with textile sectors where stringent environmental standards of developed countries have affected India's export, especially when nearly 40 per cent of the India's textiles are directed to the EU. High costs are borne by the firms to make suitable changes, particularly in the case of the small and medium enterprises, apart from lack of technical expertise and manpower.

German textile industry in 1993 introduced two types of eco-labels; the MST (*Marke scadstoffgeprufter Textilien*) that relates to attributes of the final product and the MUT (*Marke umweltschonender Textilien*) that relates to the production process of textiles. German firms were able to comply with these standards easily as they themselves were involved in development of the standards and hence already possessed the required technical competence. However, this was not the same for India or other developing countries, as they had to import dyestuffs from Germany or other EU countries, thus resulting in an exorbitant rise in the cost of certification. In addition, with respect to MUT, an on-site inspection was required, which mean further additional cost.

Germany had also banned imports of textiles and clothing using amine-based azo dyes, which are harmful to the health of textile workers. These dyes comprise 75 per cent of the dyes produced and used in India; textiles account for 30 per cent of India's exports.[16] Also the use of dyestuffs such as cobalt blue and sulphur black have been totally banned in the international market. Though workable substitutes have been explored, switching over to them entails investments of over US$13 million, mainly for the upgradation of technology and new treatment plants in order to obtain the requisite quality.[17]

3. Ecomark in India

The swift but steady rise in use of eco-labels on multiple products world over impelled India to come up with its own scheme. The Ministry of Environment and Forests launched the scheme of labelling of environment friendly products in 1991, for their easy recognition. The label, 'Ecomark', is conferred to goods that meet the specified environmental criteria and the quality requirements of Indian Standards.

The label is conferred after thorough assessment by three committees, namely Steering Committee (Ministry of Environment and Forests), Technical Committee (Central Pollution Control Board) and the Bureau of Indian Standards

(BIS), which were constituted in the very same year. The committees comprise members having specific expertise and representatives of the industry affected, the scientific community, environmentalists consumer groups and government.

The inter-ministerial Steering Committee determines the product categories to which an Ecomark may be granted and the Technical Committee determines whether the specific product needs to be included under the scheme. Simultaneously the Steering Committee is also responsible for promoting the labelling scheme to manufacturers and consumers.

While the Technical Committee is the central committee, there are also various subcommittees for the development of Ecomark criteria. After a particular product is selected to cover within the ambit of the scheme, the BIS translates the product criteria into Indian Standards.[18] They assess and certify the products, and coordinates with manufacturers wishing to use the Ecomark label on their products.

Initially the Ecomark scheme was initiated for 16 products[19] including leather and leather products in 1991. However, the scheme could not become popular, as it lacked ample response from the manufacturers. Despite, the Ministry initiating publicity campaigns for providing necessary awareness among the consumers and manufacturers on the same.

Quite a lot of reasons are there for the Indian producers not to opt the scheme. Other than mere lack of environmental awareness among the domestic market in our country, such schemes are also quite expensive being a self-financing program. Manufacturers need to pay for the costs involved in certification, i.e. cost for the application, testing, licensing fee, and renewal.

Even if spend, this mark is of no help in the export market since the EU doesn't recognize this certification though the scheme is founded on a similar "life cycle analysis" criteria and adopts environment standards corresponding to those applied internationally. All the more, the products need to comply with BIS quality standards before being able to apply for the Ecomark. Which means complying with a burden of another layer of regulation and approvals for manufacturers.

According to Dr. Sudhir K. Ghosh, Member Secretary of the Ecomark Technical Committee, "Indian industries are not coming forward to get eco-certification of their products, though they are involved in the process of criteria development." Some ascribe this to the costs involved in applying for the Ecomark and the numerous regulatory requirements manufacturers must meet before being awarded the eco-label. Other reasons may include industries' concerns about the Ecomark program, which are outlined above.[20]

4. Eco-label and Trade-Environment Disputes

As the environmental labels have trade distorting effect, there are a series of verdicts by the GATT/WTO on the same. However hardly any discussed eco-labels directly, though they do shed much light on the evolving treatment by the WTO on aspects of labelling of products. The most related disputes for eco-labels arising under the GATT system were *Malt Beverages*, *Auto Taxes*, and the two *Tuna-Dolphin* cases. Since the establishment of WTO, the most relevant disputes are *Asbestos*, *Sardines*, and the line of *Shrimp-Turtle* cases.

These disputes, brought before the GATT/WTO have highlighted the conflict faced by governments who wish to pursue domestic environmental policies and at the same time comply with WTO rules.[21] For instance, the WTO principles of non-discrimination require that like products are treated equally irrespective of the country of origin. This poses a problem for the environmental lobby since it is often the production method that differentiates an environmentally friendly commodity, from one that is not.

U.S.—Restrictions on Imports of Tuna[22]: In the much-cited Tuna-Dolphin case in 1991, Mexico used the most-favoured nation clause of Article I:1[23] GATT against the mandatory "dolphin-safe" label[24] in the US, as Mexico felt discriminated by the label. However, the panel set aside the claim, it noted that the labelling provisions of the Dolphin Protection Consumer Information Act (DPCIA) do not restrict the sale of tuna products. As these tuna products can be sold freely both with and without the "Dolphin Safe" label. Nor do these provisions establish requirements that have to be met in order to obtain an advantage from the government. Any advantage, which might possibly result from access to this label, depends on the free choice by consumers to give preference to tuna carrying the "Dolphin Safe" label. The labelling provisions therefore do not make the right to sell tuna or tuna products, nor the access to a government conferred advantage affecting the sale of tuna or tuna products, conditional upon the use of tuna harvesting methods. Thus, whether or not an eco-label contradicts the most-favoured-nation principle depends on whether or not one country favours or discriminates against another country selling the labelled product.

The Panel also commented on Article IX:1[25] as Mexico considered the labelling provisions of the DPCIA to be marking requirements falling under this Article. Here also the panel held it the other way round. The panel noted that Article IX refers to "Marks of Origin" and its text refers to marking of origin of imported products. Besides the Article does not contain a national-treatment but only a most-favoured nation requirement, which indicates that this provision is intended to regulate marking of origin of imported products but not marking of products generally. Hence the Panel held that the labelling provisions of the DPCIA do not fall under Article IX:1.

Environmental community greeted this decision on the issue of labelling by the Panel. Further holding that it would be a powerful tool for environmental policy advances besides setting an important leeway to some of the more troubling trade and environmental disputes.

However, this case cannot be held as a foreteller about the future WTO panels and the way they would interpret the TBT Agreement most favoured nation requirement. For the sole reason that the TBT Agreement[26] was not in existence when Tuna/ Dolphin I was decided.

European Communities—Trade Description of Sardines:[27] In the Sardines Dispute, the WTO's Appellate Body held that a EU regulation, which allowed only one fish species to be labelled as 'sardines', violated the WTO's Agreement on Technical Barriers to Trade. The Appellate Body made several conclusions regarding the TBT Agreement that may have important implications for eco-labelling and other regulations that define product characteristics.

First, it ruled that the obligations of the TBT Agreement are triggered whenever a Member country requires specific names to be used for marketing specific products or otherwise defines the characteristics of a product. Merely 'naming' or 'defining' a product may constitute a 'technical regulation', even if the name is based on characteristics intrinsic to the product.

Second, it also suggested that the TBT Agreement strongly favors compliance with an international standard. Both the Appellate Body and the Panel gave a hard look to the EU's argument that the international standard was an 'ineffective or inappropriate' means to fulfill the legitimate objectives of its regulation.

5. Eco-labelling in World Trade Talks

According to a news report[28] in Environment News Service, consumers might have to do away with the option of making environmentally informed choices. This situation would rise if the WTO talks succeed to ban energy efficiency and recycling labels during the ongoing negotiations. Countries including Korea, US and China are asserting that eco-labelling is acting as a barrier to trade and hence affecting their competitiveness. A proposal regarding the same, to outlaw energy efficiency labelling, is been put forward for discussion at the WTO's ongoing negotiations. If the discussions turn out to be favourable then its ultimate consequence might result in denial of consumers' access to environmental information on a wide range of products. Up till now, the consumers can choose environment friendly products, because products are labelled.

As per the news report the basic measures that countries want to be revoked include:

- labels which show whether a product is recyclable, such as the triangle logo found on plastic bottles
- labels which show whether a product, such as paper, has any recycled content

- labelling which indicates sustainable sourcing, such as dolphin-friendly tuna
- safety testing on imported foods, like compulsory testing for lethal toxins in shellfish
- tax breaks for fuel efficient vehicle engines
- fines for manufacturers, when testing shows their products fail to meet their energy efficiency claims
- labels which show the energy efficiency rating of appliances such as washing machines, fridges and irons.

The Non-Agricultural Market Access negotiations at the WTO currently looks into these measures that are being challenged, including the objection to energy efficiency labelling by Korea, as they are classified as non-tariff barriers. Many in developing countries also see eco-labelling as a non-technical barrier to trade because it is often linked to environmental agreements. However, according to Jorge Larson Guerra, a biologist with the National University of Mexico, these so-called barrier can actually help promote trade because labelling can offer a means of marketing a particular competitive advantage said.

6. Conclusion

Eco-labelling is thus an effective way to protect environment, provided it is non-discriminatory, with principles and procedures that have wide acceptance both nationally and internationally. But for attaining the same the countries need to harmonize different eco-labelling schemes, ensure mutual recognition and provide technical assistance to developing countries. So that such schemes could be made more effective, and at the same time ensure a lesser amount of negative implication to free trade.

Hence to reduce any future impact of eco-labelling on exports from developing countries certain measures need to be taken urgently. Eco-labels now needs to be more focused and effective than ever before. The duty lies not only upon the respective government but also upon the concerned industries and agencies that come up with such schemes.

The government of the respective countries needs to provide technical assistance and capacity building to their exporting companies, in order to upgrade the manufacturing facilities of these companies. Thus assisting them to adopt environmental friendly technologies, besides making the eco-labelling scheme transparent. Another important role for government is to ensure that eco-labels deliver on their claims. It is high time for governments to actively begin to regulate eco-labels.

Similar is the responsibility of the developed countries towards the developing and the least developing countries. Full-fledged assistance needs to be rendered by these rich countries for enabling the small/poorer countries to develop environmental standards that are at par with the standards set by the richer. For this reason it is necessary to build an international supporting system that would enable developing countries to produce environment-conscious products without bearing an undue burden. Similar is the need for the intellectual support for the development of environment-conscious products in developing countries.

Also the concerned agencies ought to make available the required vital information to the government and industries of developing and least developed nations. In addition, the industries in developed countries need to build long-term relationships with exporting companies. They should assist such companies in meeting the eco-label criteria.

Of the most important, the vast differences between the environmental concerns of various countries should be recognized. According to Elizabeth Barham, Dept. of Rural Sociology, University of Missouri, Columbia, the consideration of place in eco-label design is crucial for attaining the sustainability goals of labelling advocates. Place acts as a link between label principles and specific criteria, establishes a basis for accountability, and can be used as a platform for participatory processes of label creation and improvement. Place brings economic, environmental and social factors together in a way that non-place-specific or "virtual" labels cannot.

Mutual recognition of the eco-labels is also one important step to address the problem. There is a need for developing a roadmap in the direction of mutual recognition of eco-labelling schemes by going towards an increased cooperation among developed and developing countries' bodies to increase the reliability of the eco-label as a marketing instrument. Such a step towards the simplification of the eco-labelling universally will facilitate both the procedures and consumers globally.

The Indian Ministry of Commerce had once presented a paper at the "Seminar on Trade Effects of Eco-labelling" in Bangkok, Thailand, in early 1997 in an effort to achieve harmonization and mutual recognition with other eco-labelling programs. Wherein it had recommended for setting up of an organization, named the Asian Environmental Network (AEN), parallel to the Global Eco-labelling Network (GEN). The proposed AEN would be specifically covering the Asia Pacific region, with the sole purpose for better exchange and dissemination of information about eco-labelling, and to work toward greater harmonization among eco-labelling programs in Asia. Further to provide technical assistance to countries trying to develop or who are trying to set-up eco-labelling programs. However, the proposal had not been materialized.

In unison, the government decision-makers need to improve their awareness on the importance of implementing policies that encourage the diffusion and the marketing of eco-labelled products inside and outside the countries. This would improve their capability of developing sustainable procurement policies that can encourage eco-labelling diffusion and national and local policies that can support domestic industries' (particularly SMEs') efforts to have eco-labels awarded to products, such as trade export promotion, funding and investment policies.

Apart from all these what matters the most is the need for greater consumer awareness. The effectiveness of any labelling program largely depends on whether consumers actually choose to purchase the product with a label, instead of an alike one without a label. Therefore, consumer education is crucial to the success of these environmental programs. The involvement of and support by environmental NGO's (i.e. non-government organisations), consumer organisations and the media are key factors that have contributed to increasing the level of consumer awareness of environmentally preferable products in certain countries.

Box 3: How to Reduce the Trade Impact of Ecolabels

The real and potential problems faced by developing countries in relation to environmental labelling are a consumer issue. Consumers have a right to expect that the products they buy be produced in as environmentally friendly manner as possible, irrespective of source. However, consumers also have a responsibility to ensure that the prospects for advancement by developing countries are not unjustifiably hampered by a desire to promote 'greener' production. In this light the following steps can be taken now to make environmental labelling schemes more open and responsive to developing country producers:

- Outside participation and procedural openness must increase. Developing countries often do not have access to the process of choosing and applying for eco-labels. Eco-labelling bodies need to involve developing country producers whenever a product category is chosen that is of interest to them.
- Processes and methods of investigation must be standardised. The processes and methods of investigation used by environmental labelling schemes need to be standardised. Environmental labelling bodies need to follow an internationally agreed Code of Good Practice, similar to the one agreed to at the GATT under the Technical Barriers to Trade Agreement. This must involve, not just guaranteed access for developing country producers, but a guaranteed access for consumer organizations.
- Schemes must work toward the principle of 'different, but equal'. Eco-labelling schemes must come to terms with the fact that different countries use different production methods and that the one followed by their domestic industries is not necessarily the only acceptable way of operating. Work must be carried out to develop a means of labelling products manufactured under differing methods.

- Countries must recognise each other's labels. Work must be carried out to allow countries with different labels and schemes, based on different environmental criteria, to mutually recognise each other's labels.
- Developing countries must be given preferential access to 'green' technology. The transfer of 'green' technology to developing countries should be at the centre of any efforts to reform production process on a global scale.

Source: Pradeep S. Mehta & Phillip Evans, Greening Consumer Choice, Environmental Labelling and the Consumer, *Briefing Paper*, CUTS, September 1995. (Originally written for Consumers International.)

NOTES

1. Agenda 21, along with the Rio Declaration on Environment and Development, and the Statement of principles for the Sustainable Management of Forests were adopted by more than 178 Governments at the United Nations Conference on Environment and Development (UNCED) held in Rio de Janerio, Brazil, 3 to 14 June 1992. It is a comprehensive plan of action to be taken globally, nationally and locally by organizations of the United Nations System, Governments, and Major Groups in every area in which human impacts on the environment.
2. Environmental labelling is the use of labels in order to inform consumers that a labelled product is environmentally more friendly relative to other products in the same category.
3. Environmental labelling is systems for the usually voluntary granting of labels by a private or public body in order to inform consumers.
4. Introduction to Eco-labelling, Information Paper, Global Eco-labelling Network, http://www.gen.gr.jp/pdf/pub_pdf01.pdf
5. Rene Vossenaar and Roland Mollerus, "Eco-labelling and International Trade: Possible Effects on Developing Countries", *Eco-labelling and Other Environmental Quality Requirements in Textiles and Clothing —Implications for Developing Countries*, 59, International Trade Centre UNCTAD/WTO, ITC (1996).
6. Arthus Edmond Appleton, *Environmental Labelling Programmes: International Trade Law Implications* (London: Kluwer Law International, 1997).

7. Chapter 4(B)(c) provides:

(c) *Assisting individuals and households to make environmentally sound purchasing decisions.*

4.20. The recent emergence in many countries of a more environmentally conscious consumer public, combined with increased interest on the part of some industries in providing environmentally sound consumer products, is a significant development that should be encouraged. Governments and international organizations, together with the private sector, should develop criteria and methodologies for the assessment of environmental impacts and resource requirements throughout the life cycle of products and processes. Results of those assessments should be transformed into clear indicators in order to inform consumers and decision-makers.

4.21. Governments, in cooperation with industry and other relevant groups, should encourage expansion of environmental labelling and other environmentally related product information programmes designed to assist consumers to make informed choices.

4.22. They should also encourage the emergence of an informed consumer public and assist individuals and households to make environmentally informed choices by:

(a) Providing information on the consequences of consumption choices and behaviour so as to encourage demand for environmentally sound products and use of products;

(b) Making consumers aware of the health and environmental impact of products, through such means as consumer legislation and environmental labelling;

(c) Encouraging specific consumer-oriented programmes, such as recycling and deposit/refund systems.

8. *Principle 10*: Environmental issues are best handled with the participation of all concerned citizens, at the relevant level. At the national level, each individual shall have appropriate access to information concerning the environment that is held by public authorities, including information on hazardous materials and activities in their communities, and the opportunity to participate in decision-making processes. States shall facilitate and encourage public awareness and participation by making information widely available. Effective access to judicial and administrative proceedings, including redress and remedy, shall be provided.

Principle 12: States should cooperate to promote a supportive and open international economic system that would lead to economic growth and sustainable development in all countries, to better address the problems of environmental degradation. Trade policy measures for environmental purposes should not constitute a means of arbitrary or unjustifiable discrimination or a disguised restriction on international trade. Unilateral actions to deal with environmental challenges outside the jurisdiction of the importing country should be avoided. Environmental measures addressing transboundary or global environmental problems should, as far as possible, be based on an international consensus.

9. Article I (Most favoured nation clause), Article III (Non-discrimination), Article IX (Marks of origin) and Article XX (General exceptions).
10. Article XVII (National Treatment), Article XVI (Market Access) and Article III (Transparency).
11. Preamble of the Agreement, Annex 1.1 (Technical Regulation definition), Annex 1.2 (Standards definition), Article 2.2 (which states certain degree of trade restriction for environment could be justified), Annex 3 (Code of Good Practice).
12. Article 2.1 (Right to take SPS measures), Article 2.2 (Scientific evidence), Article 2.3 (Discrimination), Article 3 (Harmonisation with international standards), Article 5 (Risk assessment).
13. 2003 Global Ecolabelling Network (GEN) Annual Report: Full Text of Section 2 and 3, September 2004, http://www.gen.gr.jp/pdf/03annrpt_full_2and3.pdf.
14. The Study of the Effects of Environmental Measures on Market Access, Communication from India, Committee on Trade and Environment, WT/CTE/W/177, 27 October 2000.
15. Piotrowski Ralph and Stefan Kratz (1999), "Eco-Labelling in the Globalised Economy", IPG 4/99430 at http://www.fes.de/ipg/ipg4_99/ARTPIOTROWSKI-KRATZ.PDF.
16. Changing production and consumption patterns, Economic and Social Council, United Nations, E/CN.17/1996/5.
17. Sachin Chaturvedi and Gunjan Nagpal, "WTO and Product-Related Environmental Standards: Emerging Issues and Policy Options", *Economic and Political Weekly*, 38(1), (2003), 66-74.
18. "India Ecomark", *Environmental Labelling Issues, Policies, and Practices Worldwide*, Environmental Protection Agency, Washington, 1998, B-59.
19. Aerosols Propellants, Architectural paints and powder coatings, Batteries, Cosmetics, Electrical/Electronic goods, Food Additives, Food items, Household pesticides, Leather and Leather products, Lubricating oils, ODS free fire extinguishers, Packaging materials/packages, Paper, Plastic Products, Soaps and detergents, Textiles, Wood substitutes.

20. India's Ecomark, http://www.makingindiagreen.org/india.pdf.
21. Daniel Melser and Peter E. Robertson, "Eco-labelling and the Trade-Environment Debate", *The World Economy*, Vol. 28 (1), January 2005, 49.
22. United States—Restrictions on Imports of Tuna, Report of the Panel, (DS21/R-39S/155).
23. *Article I*: 1. With respect to customs duties and charges of any kind imposed on or in connection with importation or exportation or imposed on the international transfer of payments for imports or exports, and with respect to the method of levying such duties and charges, and with respect to all rules and formalities in connection with importation and exportation, and with respect to all matters referred to in paragraphs 2 and 4 of Article III any advantage, favour, privilege or immunity granted by any contracting party to any product originating in or destined for any other country shall be accorded immediately and unconditionally to the like product originating in or destined for the territories of all other contracting parties.
24. The Dolphin Protection Consumer Information Act (DPCIA) specifies a labelling standard for any tuna product exported from or offered for sale in the United States. Under this statute, it is a violation of Section 5 of the Federal Trade Commission Act (FTCA) for any producer, importer, exporter, distributor or seller of such tuna products to include on the label of that product the term "Dolphin Safe" or any other term falsely suggesting that the tuna contained therein was fished in a manner not harmful to dolphins, if it contains tuna harvested in either of two situations. The two situations are (1) harvesting in the Eastern Tropical Pacific Ocean by a vessel using purse-seine nets which does not meet certain specified conditions for being considered dolphin safe, and (2) harvesting on the high seas by a vessel engaged in driftnet fishing. Violations of Section 5 of the FTCA are subject to civil penalties.
25. Each contracting party shall accord to the products of the territories of other contracting parties treatment with regard to marking requirements no less favourable than the treatment accorded to like products of any third country.
26. The TBT Agreement specifically deals with voluntary standards, as well as private parties, which it regulates indirectly through their governments.
27. European Communities—Trade Description of Sardines WT/DS231/R, 29 May 2002 and WT/DS231/AB/R, 26 September 2002.
28. Challenges to Eco-labelling arise in world trade talks, http://www.ens-newswire.com/ens/oct2005/2005-10-21-04.asp.

Patterns of Regional Interaction: Lessons from East Africa

5

Nicholas Ndegwa Kimani

Introduction

The environment in Africa is deeply woven into the lives of Africa's peoples and into their traditional and cultural values, and despite the plethora of treaties, agreements and supporting international agencies, the evidence suggests continuing decline of, amongst others, biological diversity (i.e. loss of forests and of animal and plant species), massive deforestation, pollution, serious droughts and floods, and increasing water shortages (UNEP 2006). The continuing decline of environmental condition in the developing world provokes serious questions about the role and efficacy of law as an instrument for environmental management in post-colonial African societies (Richardson 2000). In the years following political independence in the 1960s, African countries, acting through the Organization of African Unity (OUA), now the African Union (AU), made repeated calls for assistance in addressing the myriad environmental problems which they face. Although over the years, African countries have made efforts to tackle their environmental problems, the results have been minimal due to among others, the dearth of expertise and institutional infrastructure providing deterrence and if possible, reversing the continued degradation. A number of steps have been taken to address their environmental problems with appropriate policy, law and institutional arrangements at the regional level. While some of the steps taken have stemmed from important regional instruments as the 1968 African Convention on the Conservation of Nature and Natural Resources, and the

1980 Lagos Plan of Action, far greater impetus has come from multilateral instruments such as the Rio Declaration and its action plan, Agenda 21, both of which emerged from the 1992 UN Conference for Environment and Development (UNCED) (Mosoti and Ravelomanantsoa 2004; Kaniaru 2001).

The paper considers a regional approach taken by three countries—Kenya, Uganda and Tanzania—to addressing their environmental problems with policy, law and institutional arrangements. The countries have been active in regional African initiatives such as the African Convention, Lagos Plan of Action as well as the Rio Declaration, and through the present regional project, administered by the United Nations Environment Programme (UNEP), have sought to operationalize those aspirations. UNEP, the first UN organ to be situated in the developing world, has made a major role in the development of environmental regimes, both in terms of developing binding treaties, as well as developing non-legally binding instruments (Freestone 1994).

The first part of this paper examines the background to the regional approach taken to the Partnership for the Development of Environmental Law in Africa (PADELIA), presently the only project dedicated to the development of environmental legislation and capacity-building in environmental law in Africa. In highlighting the effects of regulatory innovations on regional developments, section two focuses upon the treatment of environmental issues within the institutional mechanisms created under the 1999 Treaty Establishing the East African Community (EAC Treaty). The final section assesses effects of the framework environmental laws and implementing regulations on developments occurring at national developments. Through these discussions, the influence of state interests and institutional arrangements upon regional and national developments is highlighted.

I

OVERVIEW OF PADELIA

The regulatory innovations—framework environmental laws[1]—which have been enacted in Kenya, Uganda and Tanzania are the development of which was facilitated by the United Nations Environment Programme (UNEP), under a long-running project known as the Partnership for the Development of Environmental Laws in Africa (PADELIA). The intention was that having comparable laws of this nature would facilitate the undertaking of collaborative environmental activities under the auspices of the EAC Treaty. An overview of PADELIA and the EAC follows.

PADELIA is a UNEP project initiated to assist selected African countries to enhance their capacity in the development and, as necessary, harmonization of environmental laws and institutional arrangements. The overall objective of the Project is to enhance the capacity of the project countries for sustainable development through the development of environmental law and institutional capability. The first priority was to promote the understanding of various stakeholders of the structure and functioning of modern environmental law. The project operated through national task forces comprising environmental stakeholders who set their priorities to ensure that the activities were country-driven. To ensure that capacity of nationals was built; nationals of countries (project managers and consultants) were tasked with the analysis of environmental problems, existing policy and legal regimes and proposing the appropriate new legal frameworks. The reports and draft laws generated were also subjected to review by national stakeholders through consensus building workshops. The draft reports and laws are also circulated to members of the Steering Committee. To enhance institutional capacity, training workshops on diverse issues pertaining to environmental law and for diverse constituencies involved in the implementation of environmental law were held. National officers serving the project were also provided with computers, photocopiers and fax machines to facilitate their implementation of the project activities (Kameri-Mbote and Ouedraogo 2006; Doumbe'-Bille' and Sinatambou 1997).

The first phase of the project benefited seven countries: Burkina Faso, Sao Tome and Principe, Mozambique, Malawi and the East African countries of Kenya, Uganda and Tanzania. The first four countries concentrated on environmental law issues, which were of purely national interest, while the latter three countries (Kenya, Uganda and Tanzania) constituted a sub-regional project in East Africa. The first phase of PADELIA, 1994-2000, commenced with an initial grant of US$ 5 million from the Netherlands government. Phase II, 2001-2006, has yielded about US$ 5.2 million, with Belgium, Germany, Switzerland, Norway, and Luxembourg joining the Dutch to make contributions[2] (Kameri-Mbote and Ouedraogo 2006; Doumbe'-Bille' and Sinatambou 1997). Phase III is scheduled to run from January 2007 to December 2011.

There are five members of PADELIA's Steering Committee: UNEP, UNDP, IUCN, FAO, and the World Bank. UNEP, as the UN agency with primary responsibility for the development and implementation of environmental law, was designated as the institutional home of PADELIA. As UNEP does not have country-level offices, the United Nations Development Programme (UNDP) was brought on board on account of its long experience in the coordination of technical assistance and capacity building. IUCN (the International Union for the Conservation of Nature and Natural Resources, or the World Conservation Union), a hybrid organization with governmental and non-governmental membership, was also included on account of the valuable contributions to environmental conservation made by the IUCN Environmental Law Centre in Bonn. The World Bank was also included in view of its role in resource mobilization and support for the development of national environmental action plans (NEAP) in Africa. Finally, the Office of the Development Law Service of the Food and Agricultural Organization of the UN (FAO) was added, given its active role in the development of laws relating to natural resource sectors, such as land use, forestry, and fisheries. Lastly there is an independent African environmental lawyer who represents the interests of the African countries, and who sits on the Steering Committee meetings as Chair (UNEP 1997b).

The East African sub-regional project formed a special category under PADELIA, as the focus was on transboundary problems and the question of whether national laws and regulations could be harmonized.[3] In grouping together Kenya,

Uganda and Tanzania, the Steering Committee took four primary points into account. First, it was thought that as most environmental problems have a regional effect, even through the origins are national, as far as possible attempts should be made to deal with the problems through development and harmonization of laws for application on, at least, regional basis. Secondly, the Committee emphasized the significance of the common legal heritage. The three countries have legal systems based on common law traditions and a history of close interaction on legislative and judicial matters. Thirdly, their people have comparable socio-cultural traditions, which may be presumed to influence their perception of environmental problems and possible remedies. Thirdly, the three countries have a history and an experience of regional cooperation, which could be built upon by new initiatives. Fourthly, it was hoped that the East African experience could be replicated to other regions in Africa (UNEP 1997a).

Under the sub-regional component, the draft laws were developed using the following methodology. After the national processes to review existing laws, prepare draft reports and/or draft laws and their review by the national consensus building workshops, a broad based sub-regional meeting is convened. The sub-regional meeting reviews the draft reports and laws prepared and reviewed by stakeholders at national level in the three countries, proposes and agrees on areas and/or elements for harmonization on the selected environmental law issues. Based on the outcome and harmonization elements agreed at a sub-regional consensus building workshop, each project country is expected, upon return, to continue or undertake the process of preparing the draft law(s), taking into account the areas and elements for harmonization discussed and agreed at the sub-regional workshop. The process of developing and harmonizing environmental laws is intended to lead to the enactment of new and/or amendment of relevant national laws, regulations and/or administrative framework of each country. Laws and strategies for harmonization were agreed in seven priority activities: environmental standards in water air and soil; EIA regulations; hazardous wastes; forestry legislation; wildlife legislation; toxic and hazardous chemicals; and lastly legal and institutional aspects of the Lake Victoria environment, which was taken up because of the lake's character as a resource shared by all the three countries (UNEP 1998a, 1998b).

As a follow-up to the development and harmonization of the seven priority activities by the three countries, national experts recommended speedy conclusion of a regional environmental agreement. Their recommendations were reviewed by a workshop of Permanent Secretaries responsible for environment in the three countries, who concurred with the proposal for a regional treaty. Consequently they signed on 22 October 1998 a Memorandum of Understanding (MoU) for the promotion of cooperation in environmental management in East Africa. The MoU stipulated among others the need for the EAC Partner States to develop and harmonize the national framework environmental laws and the need for a system for the management of the Lake Victoria Ecosystem (EAC 2001a; UNEP 1998a).

The Lake Victoria drainage basin is the largest transboundary natural resource in East Africa, and the first collaborative and co-operative framework for the protection of the ecosystem and conservation of its resources was laid in 1994, with the adoption of the Convention for the Establishment of Lake Victoria Fisheries Organization. The objective of this convention is to foster co-operation among the Contracting Parties, harmonize national measures for the sustainable utilisation of the living resources of the lake, and the development and adoption of conservation measures. The Convention entered into force on 24th May 1996.

In 30 November 1999, the three East African States adopted the Treaty for the Establishment of the East African Community. Chapter Nineteen of the Treaty (i.e. Articles 111-114) provides for co-operation in environment and natural resources management, while Chapter Twenty (i.e., Articles 115-116) provides for co-operation in tourism and wildlife management. In addition, a number of legal instruments dealing with the management of natural resources and the environment have been negotiated, adopted and/or signed under the enabling provisions of the Treaty. For example, the Partner States signed, on 29 November 2003, the Protocol on the Sustainable

Development of the Lake Victoria Basin. Under the Protocol, the States recognize the need for sound management of the waters of the Basin in an integrated and holistic manner in order to "spur development and eradicate poverty in the Lake Victoria Basin" and agree to co-operate in several areas related to the conservation and sustainable utilization of the resources of the Basin. Among the principles adopted by the states in managing the resources of the Basin are 'equitable and reasonable utilization of the water resources', 'environmental impact assessment', and the polluter pays principle.

Two draft instruments still await adoption by the Summit of the Community. The MoU for Co-operation in Environment Management between the three States and signed on 22nd October 1998 by the respective Ministers for Environment, was saved under Article 142(1)(i) of the Treaty. That is, the three States agreed to save the document and construe it in such a way as to bring it into conformity with the Treaty. Following the decision of the 3rd Council of Ministers in November 2001, it was subsequently upgraded into a full Protocol on Environment and Natural Resources Management within the framework of the EAC Treaty. Once adopted and operational, the Protocol shall provide a stronger legal regime for sound and sustainable management of environment and natural resources of East African Community, and it will also be the base document for enactment of regional legislation on environment and natural resource management by the East African Legislative Assembly. Also awaiting adoption and implementation are the Environmental Assessment Guidelines for Shared Ecosystems in East Africa.

II
IMPACT OF LAWS ON REGIONAL DEVELOPMENTS

The treatment of environmental law and policy making within the EAC structures raised fundamental questions about how regional groupings are governed: how does it work as a decision-making system? How democratic are its decision-making procedures? Lastly, how democratic it can be? (Douglas and Wildasky 1982; Scharpf 1999; Schmitter 2000). My fieldwork revealed that national leaders had largely usurped the decision-making powers belonging to important institutions such as East African Legislative Assembly, a matter which had elicited great controversy. Only one of the two environmental protocols was in operation, the Protocol on the Sustainable Development of the Lake Victoria Basin, as it had received strong political support from the three Heads of State as well as financial backing from bilateral donors. Meanwhile the promise and potential of a bottom-up evolution of environmental governance, as expressed in the Environment and Natural Resources Management, remained unfulfilled, notwithstanding its approval in 2001 by the Council of Ministers. An examination of the institutional arrangements under the EAC Treaty thus reflected the contention within the literature that regional institutions are strongly affected by state power and interests as well as by other political forces. As a result, sometimes they will play meager roles not due to unfortunate circumstance but by design (Levy, Keohane, and Haas 1993a).

Overview of the East African Community

The three East African countries—Kenya, Tanzania and Uganda—cover a total area of 1.8 million square kilometers, and have a population of over 100 million people who share a common history, language, culture and infrastructure. They have also enjoyed a long history of co-operation under successive regional integration arrangements. These go as far back as 1917, when Kenya and Uganda formed a free trade area, which Tanzania (then known Tanganyika) later joined in 1927, thus forming a customs union. This arrangement was eventually transformed into the first East African Community in 1967. The community finally collapsed in 1977, triggering a long and acrimonious process of distributing assets and assigning liabilities to the member states.

Following the dissolution, the three states negotiated the East African Community Mediation Agreement for division of their assets and liabilities, which was then signed in 1984. Among the provisions of the Mediation Agreement, there was a stipulation that the three states explore areas for future co-operation and subsequently make solid arrangements for this co-operation. Subsequent meetings of the three Heads of State led to the signing of the Agreement for the Establishment of the Permanent Tripartite Commission for East African Co-operation on November 30, 1993. Full East African Co-operation operations started on March 14, 1996 when the Secretariat of the Permanent Tripartite Commission was launched at the Headquarters of the EAC in Arusha, Tanzania. The East African co-operation development strategy was approved one year later by the East African Heads of State, at their second Summit in Arusha on 29 April 1997. At the same Summit, the heads of state directed the Permanent Tripartite Commission to start the process of upgrading the Agreement establishing the Permanent Tripartite Commission for East African Co-operation into a Treaty. The Treaty for the Establishment of the East African Community was eventually signed in Arusha on 30 November 1999, and it entered into force on 7 July 2000. The East African Community was formally launched on 15th January 2001 (EAC 2006).

According to Article 5 of the EAC Treaty, the broad goal of the East African Community is to widen and deepen co-operation among Partner States in political, economic, social and cultural fields, research and technology, defense, security and legal and judicial affairs for their mutual benefits. Apart from the signing of the EAC treaty, another significant achievement in the EAC integration process has been the conclusion of the Protocol on the Establishment of the East African Community Customs Union in 2004. The Customs Union commenced operations on 1st January 2005, and member states have already embarked on negotiations for a common market. There are currently plans to enter into a Political Federation by 2013, which would see the inclusion of two countries—Burundi and Rwanda—in addition to the present three members (East African 2006).

Institutions of the EAC

Article 9 of the Treaty establishes the following organs: (a) the Summit; (b) the Council; (c) the Co-ordination Committee; (d) Sectoral Committees; (e) the East African Court of Justice; (f) the East African Legislative Assembly; (g) the Secretariat.

(a) *Summit*

The Summit, provided for under Articles 10-12 of the Treaty, comprises Heads of State and Government of the Partner States, and meets at least once in every year, during which decisions are made by consensus. It may hold extraordinary meetings at the request of any member of the Summit. By Article 11, the Summit gives general directions and impetus as to the development and achievement of the objectives of the Community, and reviews, amongst others, the state of peace, security and good governance within the Community and the progress achieved towards the establishment of a Political Federation of the Partner States.

The six-month campaign to rally public support for the proposed East African Federation was launched on 13 October 2006 in the three countries, during which the benefits of the federation would be explained to the region's 105 million people. During the campaign, the Heads of State would seek the views and proposals of their people on the type of Federation they would want and when they wish to have it, following which a referendum would be conducted in each of the three countries. If approved at the referendum, then East Africa will become a federation in 2013. The significance of the federation is that the region will have a common president, flag, anthem and currency. The presidency will rotate among the three presidents. The federation president will be a symbolic figure, as each country will retain her president, national flag, national anthem plus its own economy within the union. The federation will have an enlarged Parliament and will allow free movement of people and easy trade among the members (Njeru 2006).

The consensus among my informants, however, was the three East African leaders had gone too fast with respect to fast-tracking the Political Federation. In the process a host of practical legal and institutional issues such as free movement of persons, labour, services, goods and the right of establishment and residence had not been addressed. Although the Summit had discussed regional environmental issues, it was with respect to Lake Victoria, and not the Protocol on Environment and Natural Resources Management.

(b) *Council of ministers*

The Council of Ministers, provided for in Articles 13-16 of the Treaty, is the main-decision organ of the community, and consists of the Ministers responsible for regional co-operation. It meets twice in each year, although extraordinary meetings may also be held. Amongst its functions, provided for under Article 14, is to monitor implementation of the programmes of the Community and ensure the proper functioning and development of the Community in accordance with this Treaty. In this capacity the Council is empowered to, amongst others, make policy decisions; initiate and submit Bills to the Assembly; give directions to the Partner States and to all other organs and institutions of the Community other than the Summit, Court and the Assembly. Lastly, the regulations, directives and decisions of the Council are binding on the Partner States, on all organs and institutions of the Community other than the Summit, the Court and the Assembly.

It should be pointed out that there exists a conflict of legislative power between the Council and the East African Legislative Assembly, where the Council in exercising its law-making functions, would over-ride and over-shadow the Assembly (Oloo 2005; Wanyande 2005). The Council of Ministers has made a number of regional environmental policy decisions, for example upgrade the MoU on Cooperation in Environmental Management in October 1998, to a full protocol, following the decision of the 3rd Council of Ministers in November 2001. It has also approved the Protocol for Sustainable Development of Lake Victoria Basin, 2004.

(c) *Coordinating Committees and Sectoral Committees*

The Co-ordination Committee is provided for in Articles 17-19 of the Treaty, and consists of the Permanent Secretaries responsible for regional co-operation in each Partner State. It is empowered to implement decisions of the Council as the Council may direct, receive and consider reports of the Sectoral Committees and co-ordinate their activities. It may also request a Sectoral Committee to investigate any particular matter. Sectoral Committees, on the other hand, are provided for under Articles

20-22 of the Treaty. Subject to any directions the Council may give, each Sectoral Committee is responsible for the preparation of a comprehensive implementation programme and the setting out of priorities with respect to its sector.

For example, the Sectoral Committee on Environment and Natural Resources is composed of senior officials from the Partner States, who are drawn from the Ministries responsible for Water, Fisheries, Environment, Forests, Tourism and Wildlife issues. In addition, regional NGOs and representatives from the private sector are also represented in the committee. There are four working groups, which support the Sectoral Committee. These working groups are concerned with Terrestrial Ecosystems; Aquatic Ecosystems; Policy, Legal and Institutional issues; and Pollution issues (Okorut 2005).

(d) *East African Court of Justice (EACJ)*

Provided for under Articles 23-47, the EACJ's role is to administer justice and ensure the adherence to law in the interpretation and application of, and compliance with, the treaty. The Court ensures that Community law is interpreted and implemented in line with the Treaty. In addition to carrying out other functions, the Summit, the Council or a Partner State may request the Court to give an advisory opinion regarding a question of law arising from the Treaty and which affects the Community. In doing so, the EACJ provides an effective system of judicial safeguards to ensure adherence to law, thereby contributing to the effective participation of the EAC in the larger global economy (Ruhangisa 2005).

It should be pointed out that since the Court's inauguration on 30th November 2001 no case on environmental issues has yet been brought before it. However, in its first ever case, delivered on 4 October 2006, three members of the East African Legislative Assembly filed suit against the East African Community, particularly the Council of Ministers, alleging interference by the Council of Ministers with the functions of the Assembly.[4]

(e) *East African Legislative Assembly (EALA)*

Provided for under Articles 48-65 of the Treaty, the EALA was formally inaugurated on November 30, 2001 as the independent legislative arm of the Community. It meets at least once in every year at Arusha, Tanzania. There are 27 elected Members of EALA, nine from each partner state and five ex-officio Members—the three Ministers responsible for Regional Cooperation, the Secretary General and the Counsel to the Community. As provided for under Article 49, the EALA is empowered to debate and approve the budget of the Community; consider annual reports on the activities of the Community; and discuss all matters pertaining to the Community and make recommendations to the Council as it may deem necessary for implementation of the Treaty. Under Article 59, members of the EALA may introduce motions or bill in the Assembly. Once enacted the legislation of the Community is effected by means of Bills passed by the Assembly and assented to by the Heads of State as an Act of the Community.

Also regional parliaments are expected to enact legislation, build consensus on issues that are common to the region, yet contentious; provide an indication of political hurdles to regional integration, articulate varied societal interests and enhance a sense of regional identity, the consensus appears to be that EALA had not yet fully taken up this role, even though it had played an important part in regional integration, for example by enacting the EAC Customs Management Act of 2004, which paved the way for the establishment of the East African Customs Union on 1 January 2005 (Oloo 2005; Wanyande 2005). There was also a conflict of legislative power between the Council of Ministers and the EALA, leading to fears that the council may override and overshadow the EALA (Oloo 2005). This was seen in the Council of Ministers withdrawing a number of Bills from being tabled before the EALA and not permitting debates to proceed, and it eventually resulted in the matter being brought before the EACJ for determination.[5]

(f) *Secretariat*

The Secretariat is provided for in Articles 66-73 of the Treaty. As the executing organ of the EAC, the Secretariat is responsible for strategic planning, management and monitoring of programmes. Article 71 vests the Secretariat with, amongst others, responsibility for initiating, receiving and submitting recommendations to the Council, and forwarding of Bills to the Assembly through the Co-ordination Committee; the strategic planning, management and monitoring of programmes for the development of the Community; the general promotion and dissemination of information on the Community to the stakeholders, the general public and the international community; and the implementation of the decisions of the Summit and the Council. It should also be pointed out that the secretariat subscribes to the principle of 'subsidiarity'—i.e. those programmes which are best done by appropriate actors at the level of Member States continue to be done at that level and those which can best be done at the level of the EAC are undertaken at that level (EAC 2001b).

An organogram showing the main organs of the EAC is given in Appendix 1 to this paper.

III

IMPACT OF LAWS ON NATIONAL DEVELOPMENTS

The regulatory innovations at national level which are considered within this paper are the national framework environmental legislation as well as implementing regulations, which were developed under the auspices of PADELIA. In summary, framework environmental legislation lays down general principles regarding air quality, water quality, disposal of effluent and solid wastes and conservation of energy resources. The elaborate definition and regulation of these specific sectors is left to specific implementing regulations. In addition, an apex agency is created to, *inter alia*, enunciate central environmental policy, oversee the issuance and maintenance of minimum standards, intervene in enforcement procedures; promulgate standards on inter-sectoral matters; provide machinery for public participation in the management of the environment and natural resources; and determine the issue of licenses for activities affecting environment and natural resources. In execution of its mandate, the apex body is also expected to coordinate its activities with line ministries or departments (UNEP 2001; Wabunoha 2005).

It is beyond the scope of this paper to discuss the workings of framework environmental legislation and implementing regulations, but two broad themes underlie my empirical findings regarding the impacts of the framework environmental legislation on national developments. The first is that national institutional arrangements at national level are of absolute importance when determining the successful operationalization, or otherwise, of national framework environmental laws. In this context, performance of the national apex environmental agencies of Kenya, Uganda and Tanzania could be viewed in terms of the success or hurdles faced, in operationalizing the framework environmental laws. The second broad theme is that questions of environmental governance at national level are linked to wider questions of political governance within that particular country—particularly in the controversies raging around government-sanctioned excisions of gazetted forests in Kenya, allocation of forests and wetlands and dam-construction projects to foreign-investors in Uganda, and gold mining in Tanzania.

Impact of National Institutional Arrangements

The importance of national institutional arrangements is best-exemplified by Tanzania, where continued institutional gridlock and resultant paralysis in environmental and natural resource management has been the constant theme in the country's environmental management. As a result of the jurisdictional overlaps in Tanzania's environmental and natural resource legislation, abuses of power, interference in decision-making and the marginalization of the public in decision-making processes in respect of environmental resource were said to be widespread, inhibiting the development of best practices and norms in environmental law regime. This was seen in the Department of Environment (DOE), falling under the Vice President's office, "taking over" of powers and functions of the National Environmental Management Council (NEMC). This has effectively left the DOE in charge of environmental matters in Tanzania (LEAT 1999). One result has been that the donor community has been reluctant to support environmental issues in Tanzania, particularly where it meant paying in sums of money directly to organs of central government, which is unlike the case in Uganda and, to a lesser extent, Kenya.

Uganda's experiences were unique, arising, as it were, from the public sector reform processes which took place in the years following the instability of the 1980s. Public sector reforms undertaken during the late 1980s and early 1990s related to local government, environment and constitutional review. As one informant explained, by the time the 1995 constitution was passed, which contained very strong environment-related provisions, great momentum and enthusiasm for environment had been generated. Consequently, PADELIA, which had only just been launched the previous year, was seen as an opportunity to translate the aspirations expressed within legislation into action. As a result of the ongoing public restructuring, there also arose a strong culture of public consultation which resulted in a more 'entrenched' ownership of the various national legislation and policies which were produced during this period. In addition, Uganda's National Environmental Agency (NEMA-Uganda) is hailed for its high standard of professionalism among its staff members. It should also be pointed out that NEMA-Uganda's has enjoyed such generous financial support from the World Bank, from its inception in 1995 to date, that the Ugandan Government has not made any financial contribution for its staffing and operational costs. This heavy reliance on donors is obviously unsustainable. As pointed out, donor support lasts for a specific period after which the beneficiary country is required to carry on the activities on its own (Situma 2000).

Despite Kenya having a comprehensive statutory framework for managing its environment and natural resources, under which a diverse range of policy-making, management and regulatory institutions are established and given mandates, its track record in implementation and enforcement of those laws has remained extremely poor. Serious shortcomings exist in the ability to follow through on important policy initiatives, and to implement and enforce laws and regulations. This can be seen in inordinate delays experienced in finalizing policy documents, laws and regulations; in the failure to translate policy pronouncements into concrete action on the ground; and in the dearth of instances in which court or other enforcement action is taken against offenders, particularly with success. NEMA also suffers from severe financial resources, which handicap its ability to discharge its mandate. As stated in its Strategic Plan, 2005-2010, it requires the sum of KES 6.2 billion, over US$ 50 million, to undertake its nationwide mandate. However, its financial allocation for the year 2005/6 was KES 273 million, or just under US$ 4 million. The limited financial resources at its disposal has meant that although NEMA has an office in each of the country's 71 districts manned by one District Environment Officer (DEO), and an office in each of the country's eight provinces manned by one Provincial Environment Officer (PEO), who act as focal points for NEMA's coordination and regulatory role at the local level, they do not all have sufficient financial or organizational resources, such as offices or office furniture, necessary to undertake their work (Mumma 2006; Kamau 2005; NEMA-Kenya 2005).

Linkage between Environmental Governance and Political Governance

The literature contends that leaders of weakly institutionalized states may genuinely want to conform to international norms and principles and comply with regime rules, but may lack the political legitimacy or the loyalty of competent and honest bureaucracies, necessary to develop and implement domestic initiatives (Levy, Keohane, and Haas 1993b). This contention holds true in Kenya's case, for example, where the political leadership lacks the will to prioritize environmental issues, notwithstanding the presence of the 2004 Nobel Peace prize winner—Professor Wangari Maathai—a renown environmentalist (Mumma 2006). Uganda faces similar criticism, not because of its environmental laws, but because the government was nonetheless allocating protected forests and wetlands to private developers and investors, for real-estate projects, expansion of sugarcane plantations and for construction of hydro-electric dams. To one informant, there could only be one conclusion: the government either never meant to implement the environmental laws, or they never envisaged how far-reaching the environmental laws were.

The honesty of environment-related bureaucracies in Tanzania came under question, for example when it came to the sharing of information. The country was said by informants to be suffering from a legacy of authoritarianism and one-party state, with the result being an institutional culture where the Government has never felt compelled to open its files. Particular reference was made with regard to the secrecy surrounding contracts for mining gold, diamonds and natural gas, even though such contracts were binding on the government for up to 50 years. To my informants, the reason was clear: the terms of the mining contracts remained secret to ensure the political survival of the bureaucrats and politicians who would be greatly scandalized if the public came to know how they had suddenly become very rich, due to their underhand dealings. As one informant asserted:

> We tend to forget that governments are run by bureaucrats. Bureaucrats serve their own interests, first and foremost. Politicians are figureheads. Bureaucrats are the ones who rule, but they run behind their political masters. (Personal Interview, 1st September 2006)

Conclusion

The theoretical and empirical issues discussed in the foregoing paragraphs have provided the opportunity for outlining patterns of interaction between state actors and non-state actors within the context of environmental governance in East Africa. It is noted that the three countries face major challenges in managing their national and transboundary natural resources for sustainable development and poverty alleviation. Also noted is that in developing legal frameworks and related institutional arrangements for rational utilization of those resources, the three countries have applied endogenous processes and mechanisms which have been developed under the auspices of PADELIA. Through PADELIA, the three countries have developed legal instruments as well as built up expertise in environmental law and policy, which would then be expected to make effective use of those legal and institutional frameworks. A second major observation is that the development of sub-regional activities is difficult when countries are not at the same level of environmental law development and have specific national agendas that may not coincide with the aims and objectives of the sub-regional initiative (Kameri-Mbote and Ouedraogo 2006).

Although many factors and problems converge to create a situation that is not always conducive to realizing the full potential of the legal and institutional arrangements at regional and national levels, two factors remain paramount: the levels of support or interference given by the political leadership, and the integrity of formalistic institutional arrangements. The treatment of environmental law and policy-making within the EAC structures, for example raises fundamental questions about how regional groupings are governed, in terms of its workings as a decision-making system, as well as how democratic are its decision-making procedures. As much as the Protocol on Environment and Natural Resources Management builds upon the MoU For Co-operation in Environment Management between the three States and signed on 22nd October 1998, which in turn builds upon the seven priority areas identified and addressed by the East African people under the auspices of PADELIA, the three countries have nonetheless failed to adopt and implement its provisions. Similarly, the institutional shortcomings at national level, as well as controversies raging around government-sanctioned actions in each of the three countries underline the observation that questions of environmental governance are necessarily linked to wider questions of political governance.

These observations only serve to highlight the enormity of the challenges posed by rapid restructuring of domestic and international governance towards regional outlooks. They also challenge our theoretical understanding of the bodies of literature that encapsulate, analyze and discuss the shift from a world made up by sovereign nation-states to today's inter-, supra- and transnational arrangements. Particularly with regard to the literature on regions, we may well need to take a closer look at the key actors operating within non-Western regional groupings both in the context of conventional state-centric and formalistic modes of interaction, as well as in the context of 'bottom-up' non-state actor interaction, as our understanding of their agenda, coordination mechanisms, as well as their problems and limits is still far from complete.

Appendix 1: Organisation of the East African Community

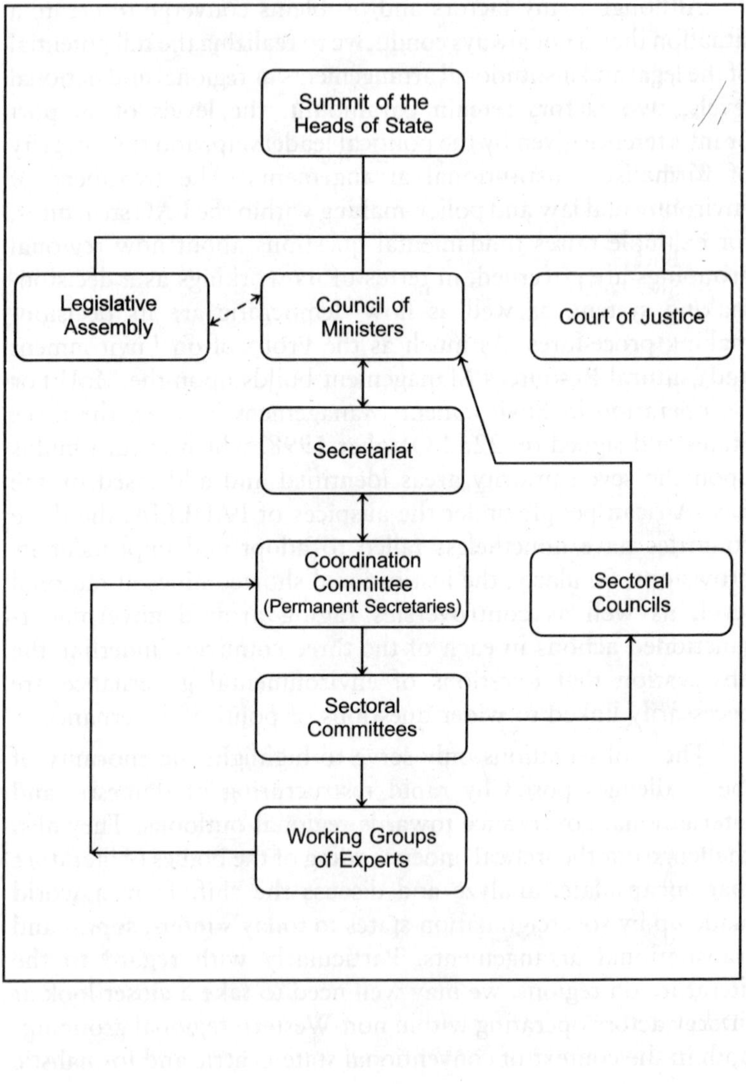

NOTES

1. An example of framework environmental legislation is Kenya's Environmental Management and Coordination Act, 1999 (EMCA). It establishes a legislative framework for the supervision, regulation and coordination all matters relating to the conservation of environment and the management natural resources in Kenya. Its scope extends to environmental conservation; the sustainable utilization, management and protection of the country's natural resources; co-ordination of actors and stakeholders in the conservation and management of natural resources and the environment; and the collection and use of information as a management tool in the natural resources conservation and management.
2. Phase of PADELIA, which is scheduled to end in December 2006, supports thirteen countries. These are grouped into three sub-regions, namely: the SAHEL sub-region comprising Mali, Niger, Senegal and Burkina Faso; the SADC Sub-region comprising Botswana, Lesotho, Swaziland and Malawi and the East African sub-region comprising Kenya, Uganda and Tanzania. Other countries involved are Mozambique and Sao Tome and Principe.
3. It is important to note that harmonization is not about having the same, or common laws or standards, but having comparable laws and principles. The activities for harmonization are tailored to suit national needs while at the same time providing a forum for sub-regional collaboration on transboundary issues, harmonization of legislation, sharing of experiences, exchange of information, and capacity building for the benefit of all project countries.
4. The judgment by the EACJ originated from a reference case, "Application No. 1 of 2005—Calist Mwatela, Lydia Wanyoto Mutende, Isaac Abraham Sepetu (Applicants) VERSUS East African Community (Respondent) (*Source*: http://www.eac.int/news 2006 10 EACJ delivers first judgment.htm) Date accessed: 14 October 2006.
5. Application No. 1 of 2005—Calist Mwatela, Lydia Wanyoto Mutende, Isaac Abraham Sepetu (Applicants) VERSUS East African Community (Respondent) where three Members of the East African Legislative Assembly alleged interference with the functions of the Assembly when it withdrew from the Assembly various Private Members Bills which had already been tabled in the House for debate.

BIBLIOGRAPHY

Douglas, Mary, and Aaron Wildasky (1982). "Risk and Culture: An Essay on the Selection of Technical and Environmental Dangers." Berkeley, CA: University of California Press.

Doumbe'-Bille', Stephane, and Etienne Sinatambou (1997). External Evaluation: UNEP/UNDP Joint Project on Environmental Law and Institutions in Africa. Nairobi: United Nations Environmental Programme.

EAC (2001a). Draft Protocol on Environment and Natural Resources Management.

—— (2001b). EAC Development Strategy: 2001-2005. Arusha: East African Community Secretariat.

—— (2006). *About the East African Community*. East African Community Secretariat 2006 [cited 11 April 2006]. Available from http://www.eac.int/about_eac.htm (date accessed: 11 April 2006).

East African (2006). "It's Official: Rwanda, Burundi Join East Africa Next Month." *East African*, October 16, 2006.

Freestone, David (1994). "The Road from Rio: International Environmental Law after the Earth Summit." *Journal of Environmental Law* 6: 193-218.

Held, David, Anthony McGrew, David Goldblatt, and Jonathan Perraton (1999). "Globalization." *Global Governance* 5: 483-96.

Jordana, Jacint, and David Levi-Faur (2004). "The Politics of Regulation in the Age of Governance." *The Politics of Regulation: Institutions and Regulatory Reforms for the Age of Governance*, ed. by J. Jordana and D. Levi-Faur. Cheltenham, UK: Edward Elgar.

Kamau, Evanson Chege (2005). "Environmental Law and Self-Management by Industries in Kenya." *Journal of Environmental Law* 17(2): 229-44.

Kameri-Mbote, Patricia, and Hubert M.G. Ouedraogo (2006). Partnership for the Development of Environmental Law and Institutions in Africa (PADELIA): Evaluation Report. Nairobi: United Nations Environmental Programme.

Kaniaru, Donald W. (2001). "Development and Implementation of Environmental Law—A Contribution by UNEP." *Environmental Policy and Law* 30 (5): 234-46.

Le Pere, Garth (2005). "The East African Community: A Theoretical Homily?" *The Making of a Region: The Revival of the East African Community*, ed. by R. Ajulu. Midrand, South Africa: Institute for Global Dialogue.

LEAT (1999). Report on Institutional Mandates and Legal Framework for Environmental Management in Tanzania. Dar es Salaam: Lawyers Environmental Action Team.

Levy, Marc A., Robert Keohane, and Peter M. Haas (1993a). "Effectiveness of International Environmental Institutions." *Institutions of Effective*

International Environmental Protection: Sources of Effective International Environmental Protection, ed. by M.A. Levy, R. Keohane and P.M. Haas. Cambridge, Mass: MIT Press.

—— (1993b). "Improving the Effectiveness of International Environmental Institutions." *Institutions of Effective International Environmental Protection: Sources of Effective International Environmental Protection*, ed. by M.A. Levy, R. Keohane and P.M. Haas. Cambridge, Mass: MIT Press.

Mittelman, James H. (1996). "Rethinking the 'New Regionalism' in the Context of Globalization." *Global Governance* 2: 189-213.

Mosoti, Victor, and Lalaina Ravelomanantsoa (2004). *An Introduction to the African Convention on the Conservation on Nature and Natural Resources*. Ed. by IUCN, IUCN Environmental Policy and Law Paper No. 56. Gland, Switzerland Cambridge, UK: IUCN.

Mumma, Albert (2006). The Place of Culture in the Enforcement of Environmental and Natural Resources Management Laws: The Case of Kenya. Paper read at 2006 Colloquium on "Implementing Environmental Legislation: The Critical Role of Enforcement and Compliance", October 16th to 20th, 2006, at Pace University, New York.

NEMA-Kenya (2005). National Environment Management Authority: Strategic Plan: 2005-2010. Nairobi: National Environment Management Authority, Kenya.

Njeru, Mugo (2006). Leaders Launch Campaign for E. African Federation. Nation, 14 October 2006.

Okorut, Tom Okia (2005). *Policy Brief on the Environment and Natural Resources Sector of the East African Community*. Arusha.

Oloo, Adams G. (2005). "The East African Legislative Assembly and the National Assemblies of Partner States: Conflict or Harmony?" *The Making of a Region: The Revival of the East African Community*, ed. by R. Ajulu. Midrand, South Africa: Institute for Global Dialogue.

Richardson, Benjamin J. (2000). Environmental Law in Postcolonial Societies: Straddling the Local-Global Institutional Spectrum. *Colorado Journal of International Environmental Law & Policy* 11 (1): 1-82.

Rosenau, James (1997). *Along the Domestic-Foreign Frontier: Exploring Governance in A Turbulent World*. Cambridge: Cambridge University Press.

Ruhangisa, John Eudes (2005). "The East African Court of Justice." *The Making of a Region: The Revival of the East African Community*, ed. by R. Ajulu. Midrand, South Africa: Institute for Global Dialogue.

Scharpf, Fritz W. (1999). *Governing in Europe: Effective and Democratic?* Oxford: Oxford University Press.

Schmitter, Philippe C. (2000). *How to Democratize the European Union—And Why Bother?* Lanham, MD: Rowman & Littlefield.

Situma, Francis D.P. (2000). "Africa's Potential Contribution to Environmental Law." *Transnational Law and Contemporary Problems* 10: 385-421.

Soderbaum, Fredrik (2004). "Modes of Regional Governance in Africa: Neoliberalism, Sovereignty Boosting and Shadow Networks." *Global Governance* 10: 419-36.

UNEP (1997a). Special Issue Bulletin of Environmental Law. Nairobi: United Nations Environment Programme.

—— (1997b). Special Issue Bulletin of Environmental Law. Nairobi: United Nations Environment Program.

—— (1998a). Report of the Meeting of the Permanent Secretaries on the Development and Harmonization of Environmental Law on Selected Topics Under the East African Sub-Regional Project. Paper read at Meeting of the Permanent Secretaries Responsible for Environmental Matters in Kenya, Uganda and Tanzania, at Nairobi, Kenya.

—— (1998b). Special Issue Bulletin of Environmental Law. Nairobi, Kenya: United Nations Environment Programme.

—— (2001). *The Making of A Framework Environmental Law in Kenya, Environmental Policy and Law Series*. Nairobi: ACTS Press.

—— (2006). African Economic Outlook 2: Our Environment, Our Wealth. Nairobi: United Nations Environmental Programme.

Wabunoha, Robert A., ed. (2005). *Handbook of Environmental Law in Uganda*. Kampala: National Environmental Management Authority, Uganda.

Wanyande, Peter (2005). "The Role of the East African Legislative Assembly." *The Making of a Region: The Revival of the East African Community*, ed. by R. Ajulu. Midrand, South Africa: Institute for Global Dialogue.

Arsenic Contamination in Groundwater of Bangladesh

6

Mohammod Aktarul Islam Chowdhury

Introduction

Arsenic, the 20th and 12th most abundant element in the earth-crust and in the biosphere, is known for its high toxicity and its ability to induce skin cancer after long-term ingestion (Driehaus et al., 1998) which is a common trace inorganic contaminant in drinking water when extracted from subsurface through tube-wells that has been identified as a significant health risk (McNeill and Edwards, 1997; Pontius et al., 1994; Abernathy and Ohanian, 1993). Possibly arsenic contamination in groundwater was first recorded as severe case in Taiwan after 1961; then the occurrences of arsenic in drinking water drawn from a variety of groundwater environment has been reported in many places around the world particularly in Argentina (Nicolli et al., 1989;), Australia (Smith et al., 2003), Bangladesh (Karim, 2000; Nickson et al., 2000; Kinniburg and Smedley, 2001; McArther et al., 2001), Cambodia (Ahmed, 2003), Chile (Ahmed, 2003), China (Inner Mongolia and Guizhou province) (Guo et al., 1997; Ahmed, 2003), Hungary (Ahmed, 2000), India (West Bengal) (Bhattacharya et al., 1997; Mandal et al., 1998), Lao PDR (Ahmed, 2003), Mexico (Arminta et al., 1997; Delrazo et al., 1990), Mongolia (Ahmed et al., 2002), Nepal (Ahmed, 2003), Pakistan (Ahmed, 2003), Switzerland (Fabian et al., 2003), Taiwan (Chen et al., 1994), Thailand (Ahmed, 2003), the USA (Korte, 1991; Welch et al., 1988, 2000) and Vietnam (Berg et al., 2001). Among them, arsenic contamination in Bangladesh has been regarded as the severest natural calamity of the world.

Extent of the Arsenic Problem in Bangladesh

In groundwater of Bangladesh arsenic was first detected in the district of Chapai Nawabgang bordering the West Bengal of India in 1993. Since then elevated levels of arsenic (exceeding the WHO guideline value of 0.01 mg/l and Bangladesh standard of 0.05 mg/l) have been identified in many regions of the country (Ahmed et al., 1997; Khan et al., 1998; Talukdar et al., 1998; Safiullah et al., 1998; Ahmed, 1998; Tanabe et at., 1998). Estimates of arsenic affected population are being updated, as more data are available. In Bangladesh, about 23 million people are "at high risk" endangered by arsenic (Dave, 1997). Data on arsenic contamination in tube well water gathered so far suggest that the presence of arsenic in groundwater is highly scattered in nature. Arsenic contamination of groundwater has been detected in 41 out of 61 administrative districts and an estimated 35 million people in Bangladesh are at risk of arsenic toxicity (Khan et al., 1998). In accordance with survey findings of Dhaka Community Hospital (DCH) published in October 1997, 60 million people of 41 districts are "at risk of arsenic contamination" as water of tube wells of 41 districts showed arsenic concentration above 0.05 mg/l (permissible limit of arsenic in drinking water standard for Bangladesh). In accordance with the study report published jointly by the School of Environmental Sciences (SOES) of Jadabpur University of India and DCH, 50 million people of 29 districts of Bangladesh are at the risk of arsenic poisoning. Field survey conducted from August '95 to February 2000 by SOES and DCH shows that 100 million people of 54 districts of area 125,133 sq. km is at risk of arsenic hazard where groundwater of 73.39% of tubewells contain arsenic above 0.01 mg/l and 9.3 million people of 47 districts of area 112,407 sq. km are at risk of arsenic poisoning where groundwater of 53.47% of tube wells contain above 0.05 mg/l (DCH and SOES, 2000). According to the recent estimate of the conducted study of 2004-2005, 27% of shallow hand-tube wells have arsenic concentration exceeding 0.05 mg/l; in acute arsenic problem areas more than 75% of shallow tube wells are contaminated with arsenic. Out of 64 administrative districts of Bangladesh, arsenic and peat are abundant throughout Bangladesh. Water quality results obtained from field study predominantly ensure the reduction theory of Arsenic release in subsurface although water level and pH data supports for oxidation theory to some extent. Thus a single mechanism is not responsible for the release and mobilization of Arsenic in subsurface of Bangladesh.

Role of Bacteria

Aerobic oxidizing bacteria

In dry season due to lowering of water table by irrigation, aerobic environment appears in the deeper sediment where organic matters can be decomposed easily by aerobic bacteria. Groundwater of many tube wells of Samta village contain significant amount of nitrogen (AAN, 1999) which is oxidized to nitrite and then soon to nitrate by available aerobic bacteria under oxic condition (Akai et al., 2001).

Anaerobic reducing bacteria

Different type of metal reducing anaerobic bacteria such as Iron and Manganese reducing bacteria are responsible to create mobilization and release of Arsenic into groundwater from different Arsenic bearing bounds. Reduction of As (V) to As (III) in anoxic environments is thought to occur primarily by dissimilatory reduction where microorganisms utilize As (V) as a terminal electron acceptor for anaerobic respiration. To date, dissimilatory arsenate reduction has been characterized in at least seven bacteria such as *Sulfurospirilium barnesii, Sulfurospirilium arsenophilum, Bacillus Arsenicoselenatis, Bacillus selenitireducens, Desulfotomaculum auripigmentum, Chrysiogenes arsenatis* and *Desulfomicrobium*. The study conducted by BUET and MIT revealed that Arsenic release is triggered by introduction of food for bacteria in aquifer. The presence of methane in water indicates that organic matter is being utilized in aquifer by anaerobic bacteria. Akai et al. (2001) suggested bacterial reductive dissolution of Arsenic caused by anaerobic condition in aquifer.

contamination has so far been reported in 61 districts. Arsenic problem alone has reduced the national safe water supply coverage by about 15 to 25 per cent.

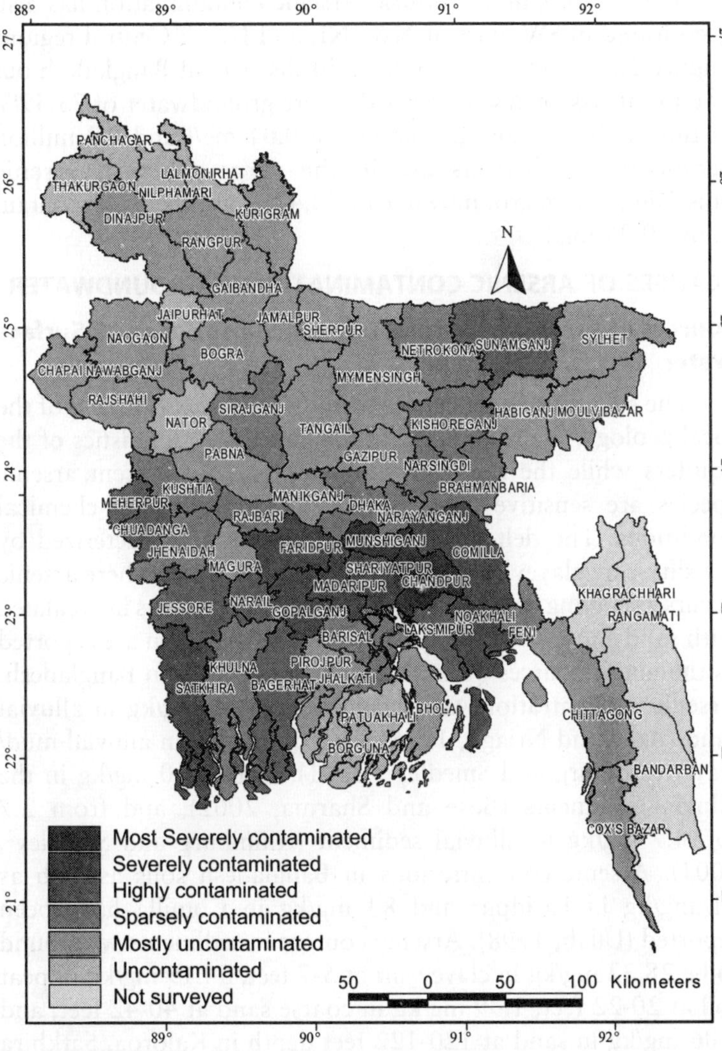

Figure 1: Severity of arsenic contamination in groundwater of Bangladesh

The most arsenic contaminated part of the country lies in the southern regions covering the districts Chandpur, Comilla, Noakhali, Lakshmipur, Munshiganj, Manikgang, Narayangang, Faridpur, Madaripur, Gopalganj, Shariatpur, Chapainawabgang, Kushtia, Jessore, and Satkhira. Arsenic contamination has also been found in SW, part of NW, NE and North Central regions (**Figure 1**). 100 million people of 54 districts of Bangladesh out of 64 is at risk of arsenic hazard where groundwater of 73.39% of tube wells contain arsenic above 0.01 mg/l and 9.3 million people of 47 districts are in the verge grasp of arsenic poisoning where groundwater of 53.47% of tube wells contain above 0.05 mg/l.

CAUSES OF ARSENIC CONTAMINATION IN GROUNDWATER

Sources of Arsenic in Natural Ground Subsurface and Surface Water

The sources of arsenic in natural water are a function of the local geology, hydrology and geochemical characteristics of the aquifers while the retention and mobility of different arsenic species are sensitive to varying geological and geo-chemical conditions. The deltaic plains of Bangladesh characterized by the silty-clay/clayey-silt as well as the sandy layers where arsenic occurs as coating on mineral grains. Clayey sediments intercalated with sandy aquifers at depths between 20 to 100 m are reported as the major sources of arsenic in groundwater. In Bangladesh, Arsenic concentration varies from 0.6 to 50 mg/kg in alluvial sand (Azcu and Nriagu, 1995), 1.0 to 6 mg/kg in alluvial mud/ clay (Kinniburg and Smedley[2], 2001); from 2-10 mg/kg in the alluvial sediments (Bose and Sharma, 2002); and from 2.7 to 14.7 mg/kg in alluvial sediment (Kinniburg and Smedley[2], 2001). Arsenic concentrations in Bangladesh soils as high as 51 mg/kg in Faridpur and 83 mg/kg in Comilla have been reported (Ullah, 1998). Arsenic content in sediments was found to be 28.32 mg/kg in clayey silt at 5-7 feet, 54.19 mg/kg in peat soil at 20-22 feet, 1.39 mg/kg in coarse sand at 40-42 feet, and 4.06 mg/kg in sand at 120-122 feet depth in Kaloroa, Satkhira district (Anwar et al., 2002). In Bangladesh, sediments appear to be typical young alluvial sediment; the average concentration of

Arsenic in alluvial sand and mud/clay have been reported to be 2.9 mg/kg and 6.5 mg/kg respectively in Bangladesh while the Arsenic content in the sediment of three specific study areas Nawabgang, Faridpur and Lakshmipur varied from 0.4 to 10.3 mg/kg with the averages of 5.9, 3.4 and 3.2 mg/kg in Nawabgang, Faridpur and Lakshmipur respectively (DPHE/BGS, 2001). Another study showed that Arsenic content varied from 1.82 mg/kg to 3.528 mg/kg in the borehole sediment of alluvial plain with the average concentration of 2.674, 2.473, 3.555, 2.55 and 3.35 mg/kg in Sylhet, Munshigang, Kushtia, Nawabgang and Chandpur were within the above mentioned typical range of Arsenic content in the natural sediments.

Datta and Subramanian (1997) reported that arsenic concentrations ranged from 1.2 mg/kg to 5.6 mg/kg in the sediments of the mighty rivers Ganges (Padma), Brahmaputra (Jamuna) and Meghna. Total Arsenic concentration varies from 1.021 mg/kg to 5.466 mg/kg in the suspended sediments collected from the rivers Padma, Meghna and Jamuna over the whole year. Arsenic content of sediment carried by the Jamuna-Padma-Meghna river system varies with particle size: 1.002 to 2.983 mg/kg in sand, 1.858 to 3.943 mg/kg in silt and 3.525 to 6.476 mg/kg in clay. Arsenic content ranged from 4.067 to 5.466 mg/kg in suspended sediments, from 2.648 to 4.736 mg/kg in river bottom sediment, from 2.462 to 5.844 mg/kg in top portion of bed sediment of flood plain and from 2.043 to 5.810 mg/kg in bottom portion of bed sediment of flood plain of the three mighty rivers.

A very large range of Arsenic concentration is found in Bangladesh groundwater from less than 0.25 mg/l to more than 2000 mg/l up to 1.460 mg/l and 2.342 mg/l in Faridpur and Chapainawabgang.

Arsenic content in surface water of the rivers even after flood is extremely low, almost negligible ranging from 0.11 ppb to 2.43 ppb mostly below minimum detection limit of the measuring equipment such as Atomic Absorption Spectrophotometer AAS.

Release Mechanism of Arsenic in Groundwater

Three available hypotheses such as *Oxidation Theory* (Mallick and Rajagopal, 1995; Mandal et al., 1998, Bose and Sharma, 2002); *Phosphate Hypothesis* (Acharya et al., 2000); and *Oxyhydroxide Theory* (Battacharya et al., 1997; Nickson et al., 1998, 2000; McArthur et al., 2001; Bose and Sharma, 2002) are in prevalence pertaining to the release and mobilization of arsenic in groundwater of Bangladesh to explain arsenic contamination.

Possibility of oxidation hypothesis

In Sylhet basin and the upper flood-plain of the Meghna, most probably arsenic carried by sediment drainage in the form of younger tertiary sediments containing pyrites from the Shillong Plateau and Lusai hills is released to the liquid phase from the sulfide mineral pyrite due to oxidation that could be increased by lowering of water table or by increasing recharge water containing dissolved oxygen.

Rejection of oxidation hypothesis

Oxidation of sulfide as the cause of arsenic release and mobilization in the sediment aquifers of Bangladesh (Chowdhury et al., 1997) gets very little authenticity. Because oxidation of sulfide minerals requires the presence of oxidizing agent and one of the key products of the reaction is dissolved SO_4; unfortunately Bangladesh groundwater containing high concentration of arsenic has generally very low SO_4 concentration typically 1 mg/l or less. Besides if sulfide reduction due to excessive withdrawal of groundwater for irrigation is regarded the cause of arsenic release in groundwater of Bangladesh, north-western part of Bangladesh would be the most arsenic contaminated zone than other area; but the reality is reverse. Even the presence of trace pyrite in Bangladesh sediment shows that it has not been oxidized and so is not a source of arsenic in Bangladesh groundwater.

Discussion on phosphate hypothesis

(a) Phosphates hypothesis can be discussed and rejected by the following arguments:

Although phosphate fertilizer such as triple super phosphate (TSP) and muriate of potash (MOP) are widely used in

Bangladesh, the amount of use are not so high in comparison to international standard on the other hand are quite below the use in the USA, Japan, China, Taiwan, etc.

(b) Other argument for rejection of arsenic-phosphate link is that phosphate fertilizer contains uranium so that this element can be used as a tracer for fertilizer phosphate. In Bangladesh groundwater concentration of uranium are mostly low being usually < 2 mg/l; where uranium is found in high concentration being slightly > 2 mg/l, phosphate concentration is very low to zero.

(c) Migration of phosphate due to surface-application of fertilizers to the aquifers is considered unlikely, at least to the deeper aquifers, considering the low mobility of phosphate in soil.

Reduction hypothesis

Organic rich sediment from Ganges delta, Padma/Jamuna/Meghna flood plain create strongly reducing condition to release Arsenic into GW by (i) reduction of strongly adsorbed Arsenic (V) to probably less strong bound Arsenic (III), leading to overall release of Arsenic, (ii) partial dissolution of Iron (III) oxide and release of Fe as well as co-precipitated and adsorbed Arsenic (& PO_4), and (iii) digenetic changes of Iron (III) oxides leading to de-sorption of adsorbed As according to Iron oxide reduction hypothesis.

Reductive dissolution of Iron oxyhydroxides and consequent release of adsorbed Arsenic could be an important mechanism of Arsenic mobilization in the subsurface as well as a major cause of Arsenic in aquatic systems in Bangladesh (Nickson et al., 2000; Kinniburgh and Smedley, 2001; Ravenscroft et al., 2001). The release of Arsenic from arseniferous Fe-oxyhydroxides by the reducing GW is most common for the sediment where rapid consumption of O_2 by microbial metabolism of organic matter produces a strongly reducing environment of the aquatic system (Lovely, 1997; Banefield et al., 1998; Chapelle, 2000; Lovely and Anderson, 2000). This process releases As with HCO_3^- and Fe^{2+} in solution as follows:

$$8FeOOH + CH_3COO^- + 15\ H_2 \Rightarrow 8Fe_2{++} + 12\ HCO_3^- + 12\ H_2O$$

Formation of farmboidal pyrite, identified by AAN (AAN, 1999) and others from Bangladesh sediment, explains the lack of

SO_4 in GW. Negative relationship between Arsenic and SO_4 in shallow aquifers ensure that GW have undergone SO_4 reduction generating reducing condition to facilitate the release of Arsenic into solution. DPHE-BGS study showed low eH in both shallow and deep aquifers down to—105 mV) ensuring the As release into GW through reduction hypothesis. Very high concentration of As from 0.012 mg/l to > 1 mg/l, dissolved Fe > 5 mg/l, Mn varying up to 2.3184 mg/l, NH_4 concentration of NH_4-N > 1 mg/l, bicarbonate > 200 mg/l higher than usual GW and PO_4 > 0.25 mg/l) but very low concentration of SO_4 < 1 mg/l and low nitrate content varied from 0.1 mg/l to < 10 mg/l further reflect reducing condition.

Generally different type of metal reducing bacteria such as Iron and Mn reducing bacteria are responsible to create mobilization and release of Arsenic into GW from different Arsenic bearing bounds. Obviously reduction of Iron oxyhydroxide (FeOOH) derived from strong Iron oxyhydroxide bounds, Sulfide bounds and strongly adsorbed Arsenic bounds (organic substance bound) and release of its sorbed Arsenic load to groundwater solution under anaerobic reducing condition is the principle mechanism that control Arsenic mobilization under reducing environment and causes Arsenic contamination in groundwater of Bengal delta including Bangladesh.

Combination of more than one hypothesis

In Bangladesh arsenic occurs in pyrite/arsenopyrite and marcasite as well as in Iron oxy-hydroxides. Many researchers discussed the sources of As and mechanism of leaching focusing on pyrite oxidation and Iron oxy-hydroxide reduction (Harvard University, 2000). Sulfide oxidation is one of the predominant processes controlling Arsenic release to GW influenced by water level fluctuations caused by GW production well pumping or climatic change. Arsenic contamination in GW may result, at least in part, from the displacement of FeOOH on sorbed Arsenic as a result of competitive anion exchange by phosphate leached from soil after excessive use of phosphate fertilizer. Reduction of FeOOH and release of sorbed Arsenic serves as a generic phenomenon for Arsenic pollution in GW aquifers where waters prevail under anaerobic condition because of organic matters

Decoupling

In decoupling phenomenon Arsenic reduction and release took place after the microbes had reduced and released Iron and other metals but not simultaneously. During decoupling Iron reduction causes a change in the mineral structure of the sediments, so that the Arsenic becomes more readily available to the metal reducing bacteria, leading to the Arsenic release into groundwater (Farhana et al., 2004). During Arsenic mobilization under anoxia, respiratory reduction of Fe (III) to Fe (II) by dissimilatory Iron reducing bacteria such as *Shewanella alga* causes coupling action with growth (Cumming et al., 1999).

Organic Stimulus

Arsenic release mediated by microorganisms such as metal-reducing bacteria is essentially stimulated by organic stimulus such as acetate, vinegar, molasses, glucose, methane, etc. that are favorite food of microbes and causes population to explode. Organic stimulus is added to stimulate an influx of organic carbon to the sediment where the microbes live that resulted in marked stimulation of Iron reduction followed by Arsenic release.

Health Effect of Arsenic Contamination

According to the reports of the national newspapers, a significant number of people are attacked by arsenic lesion of which a few die. Skin lesions such as melanosis, gangrene, hyper keratosis and hyperpigmentation in palms and soles and non-cirrhotic portal fibrosis are common arsenical lesions among the population affected by arsenic. In a later stage the patches might develop into cancer and foot will probably have to be amputated (Battacharya et al., 1997; Pontius et al., 1994). Skin lesions, skin cancer, internal cancers of bladder, kidney and lunges, neurological effects, hypertension, cardiovascular disease and diabetes mellitus, pulmonary disease and peripheral vascular disease are included to long-term health effects of exposure to arsenic (WHO, 2000). About 7,600 arsenic affected patients have been identified in arsenic affected areas. Based on upazila statistics the exposure levels to Arsenic exceeding 50 and 10 mg/l were 28.1 million and 46.4 million respectively. School of Environmental Studies, Jadavpur University, Kolkata, India and Dhaka Community Hospital estimated that the populations exposed to above 0.01 mg/l and 0.05 mg/l in 43 districts in Bangladesh were 51 million and 25 million respectively (SOES and DCH, 2000).

Effect of Arsenic Contamination in Food Chain

Food crops such as vegetables and cereals can become a path by which Arsenic may enter the food chain. Recent studies by Meharg et al. (2002) and Huq et al. (2001) demonstrate significant uptake of Arsenic by rice and a range of vegetable crops commonly grown in Bangladesh. Numerous greenhouse studies by a number of researchers have revealed that an increase in Arsenic in cultivated soils leads to an increase in the levels of Arsenic in edible vegetables. A number of studies have shown reduced growth of plants (including paddy) grown in soil containing high arsenic or when irrigated with water containing high concentration of arsenic (Abedin et al., 2002). Smith et al. (2001) has shown that yield of both tomato and silver-beet grown in soil containing 100 to 200 mg/kg arsenic, was significantly reduced (5 to 10% of the yield). High arsenic in irrigation water and soil appears to result in higher concentration of arsenic in root, stem and leaf of rice plants. Since rice straw is widely used as cattle feed in Bangladesh, high arsenic in rice stem and leaf (i.e., in straw) may result in adverse health impacts on cattle and increase human arsenic exposure via the plant-animal-human pathway. Most of the groundwater used for irrigation in Bangladesh is contaminated with arsenic (Khan et al., 1998). If arsenic contaminated water is used for irrigation, it may create hazard both in soil environment and in crop quality.

Large Scale Sludge Produced by Chemical Treatment Methods Causes Environmental Hazards

To combat the arsenic crisis, a number of household and community based arsenic removal technologies have been developed and a number of such technologies are current'y being used in many arsenic affected areas. All arsenic removal units generate some form of arsenic-rich wastes. In the absence of any clear guideline for safe disposal of wastes generated from large-scale use of arsenic removal units, such arsenic-rich wastes are disposed in the open environment and indiscriminate disposal of these wastes may lead to environmental pollution. Currently, disposal of such wastes in cow-dung bed is widely practiced; biochemical processes in cow-dung bed transform inorganic arsenic and release it into air. With increasing use of arsenic removal units, concerns have been raised regarding safe disposal of these wastes and possible contamination of environment from arsenic present in the wastes. Consequently, there is a strong demand for environmentally safe reuse of an effective disposal method for iron and arsenic contaminated sludge out of water treatment plants due to the increasing amount of sludge generated by the water treatment plants in Bangladesh. At present, there are 18 number of large scale of arsenic and Iron Treatment Plant in Bangladesh each generating about 60,000 cft arsenic rich-sludge every year (DPHE, 2002). But no proper disposal method for the highly toxic arsenic sludge waste has been developed yet. Uncontrolled disposal of the arsenic sludge may also lead to the pollution to the surface water and groundwater system. Thus the disposal of arsenic rich sludge generated from the treatment processes is one of the issues that have received little attention from the sponsors of the technologies and the users. Obviously there is a strong demand for environmentally safe reuse of arsenic contaminated sludge out of water treatment plants due to the increasing amount of sludge generated by the water treatment plants in Bangladesh.

Social Problem and Societal Aspects of Arsenic Hazard

The arsenic contamination problem has triggered a number of social and societal problems that were unanticipated and are still not fully recognized. Recent studies confirm the fact that worst arsenic problems are encountered by the poorest fraction of the society—particularly those who are already undernourished. Social fallout of arsenicosis is enormous. Among them, mental agony, social instability, superstition, ostracism, losing working ability, marriage related problems, domestic violence to the women such as dowry, divorce, separation, desertion, physical torture and polygamy are already identified as severe social consequences. Besides arsenicosis change the complexions of victims and make them physically weak, such as victim women's marital tie becomes weaker.

Economic Effect of Arsenic Hazard

Arsenicosis is more prevalent among the poor, who are forced to take contaminated tube well water and suffered with dietary deficiency due to the financial inability to buy arsenic free bottled water and nutritious food. Financial crisis also prevents to take treatment for arsenicosis. Spending huge money for treating arsenicosis increases the economic burden of middle class arsenic victim families. The poor victims become incapable of hard labor. Unfortunately arsenic victims are treated as untouchables and lose their jobs as arsenicosis acts as a barrier to have access to work. As a result arsenic victims fail to contribute to economic development of the country. Arsenic victim family loses head of household or "bread-earners" due to the disease.

Arsenic Removal Techniques Practiced in Bangladesh

The most commonly used arsenic removal technologies include oxidation, co-precipitation and adsorption onto coagulated flocs following filtration, lime treatment, adsorption onto sorptive media following filtration, ion exchange resin and membrane techniques (Cheng et al., 1994; Hering et al., 1996; Kartinen and Martin, 1995). Based on these basic techniques following arsenic removal methods under the specific categories are being employed in Bangladesh.

(a) Oxidation: Oxidation includes

- *Passive Sedimentation* (Bashi Pani: 50% reduction of As)
- *In situ Oxidation* (As reduced to half for ads. on ferric iron)
- *Chemical Oxidation* (In situ by Potassium Permanganate)
- *Solar Oxidation* (One third As reduced, Ultra violate radiation accelerate oxidation rate).

(b) Coagulation and Filtration: Coagulation followed by filtration includes:

- *Bucket Treatment Unit*

  - # DPHE-Danida Unit (Added Aluminum sulfate and Potassium Permanganate)
  - # BUET Modified Unit (Added Ferric chloride and Potassium Permanganate)
- *Stevens Institute Technology* (Iron coagulant and hypochloride)
- *Fill and Draw Units* (Community type: Oxidant and Coag.)
- *Arsenic Removal Unit Attached to Tube well* (Used in village in Bangladesh and West Bangal, India: Added Sodium hypochloride and Alum)
- *Iron-Arsenic (Fe-As) Removal Plants*
  - # Fe-As removal Unit with up-flow filter (Community type: DPHE-Danida Project)
  - # Fe-As removal plant for small town
- *Lime Treatment* (Quick lime/hydrated lime used, pH required 10.6 to 11.4).

(c) Sorptive Filtration: Sorptive filtration includes

- *Activated Alumina*
  - # BUET Activated alumina
  - # Alcan Enhanced Activated alumina
  - # Apyron Arsenic Treatment Unit
- *Granular Ferric Hydroxide*
- *Read-F Arsenic Removal Unit* (Ethylenevinyl alcohol copolymer-born hydrous cerium oxide)
- *Iron coated sand*
- *Shapla Arsenic Filter* (Iron coated brick chips)
- *Sona 3-Kalshi Filter* (1st filter: sand, iron fillings and brick chips; 2nd filter: sand, charcoal and brick chips)
- *SAFI Filter* (Porous materials like kaolinite and iron oxide)
- *Activated Carbon*
- *Indigenous Filter*

Granet Home-made Filter (Brick chips and sand)
Chari Filter (1st filter: brick chips; 2nd filter: iron fillings and sand)
Adarsha Filter (Clayey material is used)
Bijoypur Clay
Processed Cellulose Filter
Cartridge Filter (sorptive media/ion exchange resins).

(d) Ion Exchange: Ion exchange includes

- *Tetrahedron Technology* (Chlorine tablet and ion exchanger).

(e) Membrane Techniques: Membrane techniques includes

- Low-pressure Nano-filtration
- Reverse Osmosis.

A rapid Assessment of 9 household level arsenic removal technologies has become completed in Bangladesh as shown in Table 1 (BAMWSP, DFID and Water Aid, 2001). The phase-II studies of rapid assessment were conducted on 7 technologies excluding Adarsha filter and Passive sedimentation, which failed to reduce arsenic to acceptable level. On the basis of the rapid assessment, the Technical Advisory Group (TAG) of BAMWSP recommended the following household arsenic removal technologies for experimental use in arsenic affected areas:

- Alcan Enhanced Activated Alumina
- BUET Activated Alumina
- Sono 3-Kolshi Method
- Stevens Institute Technology.

WHAT TO BE DONE TO HANDLE THE ARSENIC HAZARD

Minimization with the Emphasis on Biological Removal

Not taking arsenic contaminated water is the best solution against arsenic hazard; alternative water sources such as treated surface water and rain water should be taken as the priority option in this regard specially in riverine Bangladesh boast of plenty of rainfall around 500 to 1000 mm per annum.

Table 1: Observation of rapid assessment of arsenic removal technologies

Technology	Process	Mean Flow Rate[1] (L/hr)	Mean Flow Rate[2] (L/hr)	Mean Daily Volume Achieved in Break Through Testing (L)	Average Initial Cost (US $) (1 $ = 50 Taka)	Average Initial Cost (Taka)
Alcan Enhanced Activated Alumina (AL)	Adsorption to enhanced activated Alumina	240	2875	1692	500	25,000
Adarsha Filter—An indigenous filter	Passive Sedimentation and adsorption to clay/carbonised organic candle	Excluded in Rapid Assessment due to the failure to reduce arsenic to acceptable level				
BUET Activated Alumina (BUET)	Oxidation, adsorption to sand filter and activated alumina	4.1	50	42	20	1000
DPHE/Danida 2-bucket System	Oxidation, coagulation and adsorption through sand candle.	3.6	43	42	6.5	325
GARNET Home made filter	Passive coagulation with Fe (from brick chips) and adsorption to sand	1.1	13	11	8	400
Passive Sedimentation	Passive Sedimentation	Excluded in Rapid Assessment due to the failure to reduce arsenic to acceptable level				
Sono 3-kalsi Method	Passive coagulation with Fe and/or adsorption to sand matrix	3.3	40	34		325
Stevens Institute Method	Enhanced coagulation and coprecipitation ($FeSO_4$), filtration and adsorption to sand filter	14.1	169	74	10	500
Tetrahedron	Ion exchange resin	52	624	201	240	12,000

In spite of different types of arsenic removal options especially chemically enhanced removal technologies are available to remove arsenic from groundwater extracted through tube wells, such arsenic removal systems involve large amount of sludge that unfortunately create new problem. Since in nature micro-organisms such as oxidizing or metal reducing bacteria (often algae, fungi, etc.) mediate oxidation-reduction reactions as well as arsenic release mechanism from arsenic bearing solid phase to water, mobilization and transportation of arsenic sometimes co-mobilization of both arsenic and iron/manganese occurred simultaneously by Fe(III)—reducing bacteria in the ground subsurface showing coupling action with microbial growth, extensive research activities should be initiated and emphasized against the arsenic hazard and threats following the biologically, ecologically and environment friendly remedial measures of arsenic contamination as a prevention techniques or mitigation/ removal processes using biological plant or microbes such as phyto-plankton, water hyacinth, paddy ash, etc. instead of chemical methods to avoid the sludge handling problem.

In order to minimize the intensity of arsenic hazard in food chain of Bangladesh, the rice of higher arsenic content can be blended with rice of lower arsenic content to reduce arsenic ingestion to human food chain through rice, staple food of Bangladesh. Such type of blending should be thought for other crops, vegetables, fruits, etc.

Treatment of Arsenicosis

There is a certain social stigma associated with people affected by arsenicosis, with the disease wrongly attributed to sins in the current or past lifetimes. To get rid of such undesired challenges, treatment of arsenicosis is, of course, essential. In this regard, Chelation therapy, vitamins and nutritious diet enhance the recovery of arsenicosis. Besides keratolytic oinment and cryosurgery can be applied against keratosis of palms and soles; also treatment of fungal infection with ointment and medicine can also be used in such cases.

Social Awareness

It needs to be re-emphasized the raising public awareness on the arsenic pollution and related health problems is largely a societal problem. One has to overcome the obstacles mentioned earlier to make a public awareness campaign successful. Therefore, it is critical to fully engage the civil society in developing such campaigns and in implementing them. Numerous NGOs active in this area could be an invaluable asset in this respect.

Conclusion

Bangladesh achieved a remarkable success by providing 97% of the rural population with tube well water, which is the highest population coverage by tube well water in the region. Unfortunately, recently discovered arsenic contamination of shallow aquifers in many parts of the country has made shallow tube well water unsafe for drinking. The success achieved for years in rural water supply in Bangladesh is on the verge of collapse. The problem of arsenic contamination has been complicated by a large variability at both the local and regional scale. The distribution of arsenic affected tube wells and changes in arsenic concentrations cannot be predicted even within a small area. Besides, large-scale use of groundwater for irrigation began with the "green revolution", which was initiated 30 years ago. The production of rice increased significantly and this was mainly due to the increase in rice production during dry season with the use of groundwater irrigation. But in recent years, arsenic in crops produced by arsenic contaminated irrigation water is a great concern. Thus the arsenic pollution crisis in Bangladesh is indeed severe and is pushing the limits of our knowledge and the capacity to respond to it. Considering the various dimensions of the problem discussed, it is essential to develop a cohesive strategy on a region- or nation-wide basis based on the key elements summarized as follows:

- *Raising Public Awareness* should be the kick off point for any approach to handle arsenic problem;
- *Information Dissemination and Transparency* should play a key role in effectiveness of remedial strategies

while building the confidence of stakeholders, particularly general public;

- *Participation of the Civil Society* has to be a core element of designing, planning and implementing strategies;
- *Improving Nutrition and Fighting Undernourishment* to fight against arsenicosis;
- *Alternative Livelihoods* should be provided for those who are directly impacted by arsenic contamination as well as their immediate families;
  - *Detailed investigation on source identification should be conducted;*
  - *Exploring for alternative safe water sources such as surface water sources like river, lake and ponds as well as rainwater harvesting should be kept in mind;*
- *Scientific Research* has to be emphasized with due consideration to local conditions;
- *Drinking Water Standards* have to be modified in view of the scientific research;
- *Water Resource Management* on a regional or national scale is essential to fully exploit the abundant surface water resources available—this has to include treated groundwater as a critical component;
- *Treatment Technologies* that are cheap, efficient and easy to use should be applied at a large scale as an interim or mid-term solution; Biological removal options should be taken into consideration to avoid sludge problem in stead of chemical treatment; and
- *Intra-governmental Coordination* is a must for effective implementation of the approach elements.

REFERENCES

AAN, (1999). Arsenic Contamination of Groundwater in Bangladesh. Interim Report of the Research at Samta Village. Asia Arsenic Network.

Abedin, M.J., Cresser, M.S., Meharg, A.A., Feldmann, J. and Cotter Howells, J. (2002). "Arsenic Accumulation and Metabolism in Rice", *Env. Sci. Technol.,* 36(5): 962-68.

Abernathy, C.O. and Ohanian, E.V. (1993). "Health Effects of Inorganic Arsenic in Drinking Water." *Proceeding, 1993 Water Quality Control Conference*, American Water Works Association, Denver, Colorado, USA.

Acharyya, S.K., Lahiri, S., Raymahashay, B.C. and Bhowmik, A. (2000). "Arsenic Toxicity of Groundwater in Parts of Bengal Basin in India and Bangladesh: The Role of Quaternary Stratigraphy and Holocene Sea-level Fluctuation." *Environmental Geology,* 39 (10): 1127-34.

Ahmed, M.F., Hossain, M.D., Badruzzaman, A.B.M., Jalil, M.A. and Ali, M.A. (1997). "Study on Concentration of Arsenic in Groundwater of North-eastern Zone of Bangladesh." *Journal of Civil Engineering,* Institute of Engineers, Bangladesh.

Ahmed, K.M., Nickson, R., Burgess, W.G., Macarthur, J.M., Imam, M.B., Alam, M., Ravenscroft, P. and Rahman, M. (1998). "On the Release Mechanism of Arsenic in Groundwater of Bangladesh." *Proceeding of the International Conference on Arsenic Pollution of Groundwater in Bangladesh: Causes, Effects and Remedies*, 8-12 February, Dhaka, Bangladesh.

Ahmed, M.F. (2000). "Soil Environment and Arsenic Contamination of Groundwater in Bangladesh." *Proceeding of International Conference on Geoenvironment 2000*, Sultan Qabus University, Muscat, Oman, March 4-7, 2000. Vol. 1, 256-64.

Ahmed, M.F., Minnatullah, M.M., Shamsuddin, A.J. and Ahmed, S.A. (2002). "Alternate Water Supply Options for Arsenic Affected Areas of Bangladesh." *Arsenic Mitigation in Bangladesh, An Outcome of International Workshop on Arsenic Mitigation in Bangladesh,* edited by M. Feroze Ahmed and Chowdhury Mufad Ahmed organized by Local Government Division, Ministry of LGRD and Co-operatives, Government of Bangladesh.

Ahmed, M.F. (2003). "Arsenic Contamination of Groundwater in Bangladesh." *Arsenic Contamination: Bangladesh Perspective,* ed. by M. Feroze Ahmed and published by ITN-Bangladesh.

Akai, J., Izumi, K. and Fukuhara, H. (2001). "Arsenic Release Experiments Using Strata Model—Bacterial Activity as a Key Factor for Arsenic Dissolution." *Arsenic Contamiantion of Groundwater-Geological, Hydrogeological, Medical and Sociological Findings in Bangladesh and Inner Mongolia,* Asia Arsenic Network and Research Group for Applied Geology, Japan.

Anwar, H.M., Akai, J., Mostofa, K.M.G., Safiullah, S. and S.M. (2002). "Arsenic Poisoning in Groundwater: Health Risk and Geochemical Sources in Bangladesh." *Environment International,* 27: 597-604.

Arminta, M.A., Rodriguez, R. and Cruz, O. (1997). "Arsenic Content in Hair of People Exposed to Natural Arsenic Polluted Groundwater at Zimapan, Mexico." *Bulletin on Environ. Contam. Toxicol.* 9: 583-89.

Azcu, J.M. and Nriago, J.O. (1995). "Impact of Abandoned Mine Tailings on the Arsenic Concentrations in Moira Lake, Ontario." *Journal of Geochemical Exploration*, 52, 81-89.

Banefield, J.F., Nealson, K.H. and Lovely, D.R. (1998). "Geomicrobiology: Interaction between Microbes and Minerals." *Science* 280: 54-55.

Battacharaya, P., Chatterjee, D. and Jacks, G. (1997). "Occurrence of Arsenic-contaminated Groundwater in Alluvial Aquifers from Delta Plains, Eastern India: Options for Safe Drinking Water Supply." *Water Resources Development*, 13(1): 79-92.

Berg, M., Tran, H.C., Nguyen, T.C., Pham, H.V., Schertenleib, R. and Giger, W. (2001). "Arsenic Contamination of Groundwater and Drinking Water in Vietnam: A Human Threat." *Environ. Sci. Technol.* 35: 2621-26.

Bose, P. and Sharma, A. (2002). "Role of Iron in Controlling Speciation and Mobilization of Arsenic in Subsurface Environment." *Water Research*, 36: 4916-26.

Chapelle, F.H. (2000). "The Significance of Microbial Processes in Hydrogeology and Geochemistry." *Journal of Hydrogeology*, 8: 41-46.

Chen, S.I., Dzeng, S.R., Yang, M.H., Chlu, K.H., Shich, G.M. and Wal, C.M. (1994). "Arsenic Species in Groundwaters of the Blackfoot Disease areas, Taiwan." *Journal of Environmental Science & Technology,* 28, 877-81.

Chowdhury, T.R., Manal, B.K., Samanta, G., Basu, G.K., Chowdhury, P.P., Chanda, C.R., Karan, N.K., Lodh, D., Dhar, R.K., Das, D., Saha, K.C. and Chakraborti, D. (1997). "Arsenic in Groundwater in Six Districts of West Bengal, India, the Biggest Arsenic Calamity in World, the Status Report up to August 1995." *Arsenic Exposure and Health Effects*, ed. by Chappell W.R., Abernathy C.O. Calderon R.L. and published by Elsevier Science Ltd., 93-111.

Cummings, D.E., Spring, S., and Rosenzweig, R.F. (2002). "The Ecology of Iron-reducing Bacteria in Pristine and Contaminated Environments." *Manual of Environmental Microbiology* ed. by Christon J. Hurst.

Dave, J.M. (1997). Excerpts from "Arsenic Contamination of Drinking water in Bangladesh." WHO Consultation on Arsenic in drinking water and resulting toxicity in India and Bangladesh, 5. WHO New Delhi, 29th April-1st May.

Del Razo, L.M., Arellano, M.A. and Cebrian, M.E. (1990). "The Oxidation State of Arsenic in Well Water from a Chronic Arsenicism Area of Northern Mexico." *Environmental Pollution* 64: 143-53.

Driehaus, W., Jekel, M. and Hilderbrandt, U. (1998). "Granular Ferric Hydroxide—A New Adsorbent for the Removal of Arsenic from Natural Water." J Water SRT—AQUA 47(1): 30-35.

Fabian, D., Zhou, Z., Wehrli, B. and Friedl, G. (2003). "Diagenetic Cycling of Arsenic in the Sediments of Eutrophic Bladeggersee, Switzerland." *Applied Geochemistry*, 18: 1497-506.

Farhana, S.I., Gault, A.G., Boothman, C., Polya, D.A., Charnock, J.M., Chatterjee, D. and Lloyd, J.R. (2004). "Role of Metal Reducing Bacteria in Arsenic Release from Bengal Delta Sediments." *Nature*, 430, 68-71.

Guo, H.R., Chiang, H.S., Hu, H., Lipsitz, S.R. and Monson, R.R. (1997). "Arsenic in Drinking Water and Incidence of Urinary Cancers." *Epidemio*, 8: 545-50.

Harvard University, USA (2000). Arsenic website maintained by http: phys4.Harvard.edu/wilson/Arsenic.html

Huq, S.M.I., Ahmed, K.M., Suktana, N. and Naidu, R. (2001). "Extensive Arsenic Contamination of Groundwater and Soils of Bangladesh." Arsenic in the Asia Pacific Region Workshop, Adelaide, Australia, Managing Arsenic for The Future: 94-96.

Karim, M. (2000). "Arsenic in Groundwater and Health Problems in Bangladesh." *Water Resource*, 34: 304-10.

Khan, A.A., Imam, B., Akhter, S.H., Hasan, M.A. and Ahmed, K.M.U. (1998). "Subsurface Investigation of Arsenic Contaminated Areas of Rajarampur, Chanlai and Baragharia of Nawabganj District." Research Report of Geohazard Research Group, Department of Geology, University of Dhaka, Dhaka, Bangladesh.

Khan, A.A., Akhter, S.H. and Alam, S.M.M. 2000. "Evidence of Holocene Transgression, Dolomitization and the Source of Arsenic in the Bengal Delta." A.M.O. Mohamed and K.I. Al Hosani (eds.) *Geoengineering in Arid Lands*, A.A. Balkema Publ., The Netherlands, 351-55.

Kinniburg, D.G. and Smedley, P.L.[1] (2001). *Arsenic Contamination of Groundwater in Bangladesh*, Vol. 1: Summary published by DPHE/ DFID/BGS.

Kinniburg, D.G. and Smedley, P.L.[2] (2001). *Arsenic Contamination of Groundwater in Bangladesh*, Vol. 2: Final Report published by DPHE/DFID/BGS.

Kinniburg, D.G. and Smedley, P.L.[3] (2001). *Arsenic Contamination of Groundwater in Bangladesh,* Vol. 3: Hydrochemical Atlas published by DPHE/DFID/BGS.

Korte, N. (1991). "Naturally Occurring Arsenic in Groundwater of the Mid-western United States." *Environmental Geology and Water Science,* Vol. 18(2): 137-41.

Lovley, D.R. (1997). "Microbial Fe(III) Reduction in Subsurface Environment." *FEMS Microbial Review*, 30: 305-13.

Lovley, D.R. and Anderson, R.T. (2000). "Influence of Dissimilatory Metal Reduction on the Fate of Organic and Metal Contaminants in the Subsurface." *Journal of Hydrogeology*, 8: 77-78.

Mandal, B.K., Chowdhury, T.R., Samanta, G., Mukherjee, D.P., Chanda, C.R., Saha, K.C., Chakraborti, D. (1998). "Impact of Safe Water for Drinking and Cooking on Five Arsenic Affected Families for 2 Years in West Bengal, India." *Science of Total Environment*, 218: 185-201.

McArther, J.M., Ravenscroft, P., Safiullah, S. and Thirlwall, M. (2001). "Arsenic in Groundwater: Testing Pollution Mechanism for Sedimentary Aquifers in Bangladesh." *Water Resources Research*, 37(1): 109-17.

McNeill, S.L. and Edwards, M. (1997). "Arsenic Removal during Precipitative Softening." *Journal of Environmental Engineering*, ASCE; May, 453-60.

Meharg, A.A., Abedin, M.J., Rahman, M.M., Feldmann, J., Cotter Howells, J. and Cresser, M.S. (2001). "Arsenic Uptake and Metabolism in Bangladesh Rice Varieties." Arsenic in the Asia Pacific Region Workshop, Adelaide, Australia, Managing Arsenic for the Future: 45-46.

Nickson, R., Burgess, W.G., Macarthur, J.M., Ahmed, K.M., Ravenscroft, P. and Rahman, M. (1998). "Arsenic Poisoning in Bangladesh Groundwater." *Nature*, London, UK, Vol. 395: 338.

Nickson, R.T., McArther, J.M., Ravenscroft, P., Burgess, W.G. and Ahmed, K.M. (2000). "Mechanism of Arsenic Release to Groundwater of Bangladesh and West Bengal." *Applied Geochemistry*, 15: 403-13.

Nicolli, H.B., Suriano, J.M., Peral, M.A.G., Ferpozzi, L.H. and Baleani, O.A. (1989). "Groundwater Contamination with Arsenic and other Trace Elements in the Area of Pampa, Province of Cordova, Argentina." *Environmental Geol. Water Science*, 14: 3-6.

Pontius, F.W., Brown, K.G. and Chen, C.J. (1994). "Health implications of Arsenic in Drinking Water." *Journal of AWWA*, 86(9): 52-63.

Ravenscroft, P., McArther, J.M. and Hoque, B.A. (2001). "Geochemicals and Palaeohydrological Controls on Pollution of Groundwater by Arsenic." *Arsenic Exposure and Health Effects IV*, ed. by Chsappell, W.R., Abernathy, C.O., Calderon, R.L. and published by Elsevier Science Ltd., 53-79.

Smith, L.H., Smith, J. and Naidu, R. (2001). "Assessment of Soil-plant Transfer of Arsenic from Agricultural Soils." Arsenic in the Asia Pacific Region Workshop, Adelaide, Australia, Managing Arsenic for The Future: 156-57.

Smith, J.V.S., Jankowski, J. and Sammut, J. (2003). "Vertical Distribution of As (III) and As (V) in a Coastal Sandy Aquifer: Factors Controlling the Concentration and Speciation of Arsenic in the Stuarts Point Groundwater System, Northern New South Wales, Australian." *Applied Geochemistry*, 18: 1479-96.

Safiullah, S., Sarker, A.K., Zahid, A., Islam, M.Z. and Haider, S.Z. (1998). "Geochemical Mapping and Speciation of Arsenic in the Groundwater of Faridpur Municipality." Proceeding of the International Conference on Arsenic Pollution of Groundwater in Bangladesh: Causes, Effects and Remedies, 8-12 February, Dhaka, Bangladesh.

SOES and DCH (2000). "Groundwater Arsenic Contamination in Bangladesh." Summary of Survey Reports from August 1995 to February 2000, School of Environmental Studies, Jadabpur University, India and Dhaka Community Hospital, Bangladesh.

Talukder, S.A., Chatterjee, A., Zheng, J. and Kosmus, W. (1998). "Studies of Drinking Water Quality from Chapai Nawabgang, Western Bangladesh", Proceeding of the International Conference on Arsenic Pollution of Groundwater in Bangladesh: Causes, Effects and Remedies, 8-12 February, Dhaka, Bangladesh.

Tanabe, K., Akiyoshi, Y., Yakota, Y., Hironaka, H. et al. (1998). "Arsenic Concentration of Groundwater of Samta Village and the Applicability of a Field Kit by Hironaka to Quantify Arsenic." Proceeding of the International Conference on Arsenic Pollution of Groundwater in Bangladesh: Causes, Effects and Remedies, 8-12 February, Dhaka, Bangladesh.

Ullah, S.M. (1998). "Arsenic Contamination of Groundwater in Irrigated Soils of Bangladesh." International Conference on Arsenic Pollution of Groundwater in Bangladesh: Causes, Effects and Remedies, 8-12 February, Dhaka, Bangladesh.

Welch, A.H., Lico, M.S. and Hughes, J.L. (1988). "Arsenic in Groundwater of the Western United States." *Ground Water*, 26: 333-37.

Welch, A.H., Westjohn, D.B., Helsel, D.R. and Wanty, R.B. (2000). "Arsenic in the Groundwater of the United States: Occurrence and Geochemistry", *Ground Water*, 38: 589-604.

The Troika of Environment, Livelihood and Women: Relationships and Interactions of Tribal Women in Orissa

7

Bhabani Shankar Nayak and *Geeta Sinha*

Introduction

The 2001 Census figures reveal that eight per cent of the nation's total population, over 84 million people are tribal people (Scheduled Tribes)[1] and among them 42.02 per cent of the tribal populations were main workers of whom 54.50 per cent were cultivators and 32.69 per cent agricultural labourers. Thus, about 87 per cent of the main workers from these communities were engaged in primary sector activities. The tribal agriculture and other primary sector activities like hunting, food gathering and collecting minor forest products depends on the forests. *The National Tribal Policy (NTP) 2005* says that 'as per the Forest Survey of India Report, 2003, about 60.04 per cent of the 63 per cent forest covers in the country and 63 per cent of the dense forest lie in the 187 tribal districts, though the geographical area of these districts is just 33.6 per cent of the country's geographical area. Out of 58 districts, which have more than 67 per cent of their area under forest cover, 51 happen to be tribal districts. A comparison of the 2001 and 2003 assessments of forest cover in the tribal districts shows a net increase of 321,100 hectares underscoring the point of a very symbiotic relationship between tribal, forest and its conservation' (NTP, 2005: 6-7). This indicates that with few exceptions, the majority of the tribals are forest dwellers. In spite of many competing demands and pressures on forests, tribals continue to depend on forests for their livelihood. Historically, in India, tribal perspectives on forests and government policies affecting forests have followed very different trajectories. State policies driven by industrial demands were in direct conflict with the customary use of forests by forest dwelling communities. Before state intervention, forests were managed as communal property; the crucial role of forests in the economic subsistence of individuals, families and community was the basis for managing them as communal resources (Chopra et al., 1990). A failure to recognize community control of forests led to a collapse in institutional norms that were instrumental in protecting and managing forest resources for local use. A shift in property rights to the state steadily undermined the rights of tribals to use and extract forest resources.

The colonial state policies since 1894 and post-independence forest policies have characterized tribals as indiscriminate users of forests and therefore sought to protect forests from the tribals. "Despite the policy thrusts and legislative instruments to curtail the use of forests by the local people, forest dwellers continued to find access to forests. This was due to the overwhelming dependence of the forest dwellers which make it difficult to fully administer the laws" (Pathak, 1994: 27-28). The postcolonial Indian governments have extended the same colonial forest policy (the Indian Forest Act of 1865, 1927) denying the customary rights of the tribal people over the forest and natural resources reflected in the National Forest Policy of 1952 (Guha, 1983, and Arora, 1994). Both the colonial and postcolonial governments in India have consolidated the monopolistic right of the state over forest and natural resources in the name of public benefit. Thus, historically, Indian forest policies have alienated people from the forests, thereby, exacerbating the rates of deforestation. Post-independence forest policies contributed to an expansion in industrial and mining projects to meet industrial demand for raw materials, and tightened control of forest lands through restricted access to forests and forest products (Haeuber, 1993). Protection policies increased the hardships of vulnerable social groups by denying them access to forests (Barraclough and Ghimire, 1995). But the Schedule Tribes (Recognition of Forest Rights) Bill 2005 of the Government of India recognises the rights of the tribals over forest and its natural resources but the industrialization and mining projects always overshadows the visions enshrines in the Forest Rights Bill 2005.

However, the news of eviction of rural and tribal people from their traditional habitats in favour of ongoing hydro-electric dam construction, mining and industrialization projects is the topic of regular debates in media. Its outcome is nothing but the painful experience of being a refugee in one's own home land, which has been customarily owned for generations together. The rural and tribal communities are being not only alienated from their home land but also from their natural habitat which was a source of livelihood for them. The development paths followed by the postcolonial government of Indian states are based increasingly on mono-centric industrialisation in urban centres where resources and services are increasingly concentrated whereas the rural communities and tribals in general and tribal women in particular are especially vulnerable to the processes of impoverishment. These include the degradation, depletion or loss of the natural resource base of their livelihood. While the debate goes on, there is a steep rise in the extend of degraded forest land of about 174 million hectares, threatening the subsistence of forest communities, flooding, soil erosion and heavy siltation in dams (Arora, 1994: 691). The impact is enormous when it comes to the tribal women whose livelihood depends on access to natural resources, which are the factors of production. Thus deprivation of tribal women's access to natural resources like forest is the basis of poverty generation and putting tribal women in perpetual poverty.

This study focuses on the dependence of tribals on forests in one of the Eastern Indian state called Orissa. Particular attention is given to the gender dimensions of forest use in tribal villages and the implications of industrialization on their livelihood. The present phase of industrialization and mining when viewed from a gender perspective, have significant consequences for women in tribal communities. Women are the linchpin that connects the livelihood strategies of tribal households with forest wealth. Tribal women are largely responsible for collecting and processing many of the forest products. In this way the study addresses several interrelated issues of environment,livelihood and tribal women.

This paper focuses on the relationship between the livelihood pursuits of tribal women and the surrounding natural resources like the forest, water, land etc. in Orissa. Orissa is rich in mineral resources like bauxite and iron-ore which attracts lots of industries. 'Development activities in Orissa gained momentum in the early 1950s. Officially, 81,176 families from 1,446 villages were displaced due to development projects from 1950 to 1993 that required the acquisition of 14,82,626 acres of land. The major development projects which induced large-scale displacement in the state during this period were industrial projects—such as Rourkela Steel Plant (RSP), the Hindustan Aeronautics (HAL) and the National Aluminium Company (NALCO)' (Jena, 2006). The aluminium industry is at the centre of India's new vision of itself as a budding superpower, as laid out in a book written by the India's President, India 2020 (Kalam, 1998). But this industry has a very dark side which very few people are aware of, destroying and polluting some of the world's best mountains, rivers and forests, and dispossessing thousands of tribal people (indigenous) from their land, livelihood and sustainable lifestyle (Wysham, 2001). The first part of this paper analyses the theoretical underpinnings of environment, women and livelihood issues of tribal women in Orissa. It deals with the interrelationships and different dynamics that influence the relationship between environment and livelihood of the tribal women in Orissa. Then the second part of this paper tries to analyse the present industrialization process which seems to have been violating all environmental and other regulations of the government in the form of deforestation, mining, displacement by its own institutional set ups. While locating the industrialization process, this chapter encounters the dilemmas of tribal development where the livelihood of the tribal people and women in particular are threatened. In this part of the paper certain case studies of industrialization process will be taken into consideration to analyse the existing situation. And the final part of the paper will analyse the implications of industrialization on tribal women and their livelihoods where women have been the worst victims of it and will even look at the role of state and policies related to it.

Statement of the Research Problem

The research is looking at one of the major problem faced by the tribal women in the process of industrialization in Orissa. The industrialization based on mining project is not only a threat to environment but also a predicament to the livelihood of the tribal household and the status of the tribal women. Due to the fact that tribal women are more dependent on environment which is the source of livelihood and at the same time gives a superior position to the tribal women in the tribal society. This work is trying to document the nature of change that took place in the tribal society due to industrialization from a gender perspective. It is also making a note that how industrialization is a threat to tribal women, livelihood and their position in their society.

Objective of the Study

The general objective of this study is to examine the impact of industrialization on environment and tribal livelihood. The specific objective of the study is to situate the problem of the tribal women and their forest based livelihood due to the rapid industrialization and mining projects in Orissa. In a way this study is trying to analyze the changing equations of tribal communities with the natural resources and their access to it. It intends to understand the extent to which environmental degradation has affected the lives of tribal women. The study is also trying to map out the implications of this on tribal women in terms of their identity, culture and existence in terms of livelihood issues.

Methodology of the Study

This piece of work has tried to derive the methodology of study from the 'Platform for Action' on Women and Environment that calls upon governments at all levels, international organisations, NGOs and private sector institutions, (1), to involve women actively in environmental decision-making at all levels, (2) to integrate gender concerns and perspectives in policies and programs for sustainable development, and (3) strengthen or establish mechanisms at the national, regional and international levels to assess the impact of development and environmental policies on women (United Nations, 1995). In order to cover the above tenets in the study, it has also tried to follow a process of policy-relevant theoretical development based on 'Women, Development and Environment' (WDE) framework and livelihoods approach. The work is based on secondary data collected from published sources that includes articles, journals, reports and books. The data analysis follows both qualitative and quantitative methods. The news reports and policy documents were analysed based on the method of content analysis.

Significance of the Study

The study carries an immense importance due to the fact that it looks at the most marginalized communities in Orissa, India i.e. tribals. Again the study is focusing at the issues of life and livelihood of the tribal women who are more vulnerable in the process of development projects. In terms of development of these marginalized communities with a special focus on tribal women, the study puts itself in a unique position to find out alternative public policy.

PART-1

Women, Environment and Livelihood Relationships with a Special Reference to Tribal Women in Orissa

The symbiotic relationship between environment and women is being recognised both in national and international fora. The Rio Declaration-1992 for example, recognizes women's close linkage with the environment and the role that environment play for their daily livelihood and existence. Their families too are dependent on environment (forest, land and water) for their existence. Any threat to the environment is a threat the women as they face immediate crisis in their livelihood management for themselves and their families. According to Agenda 21 of the Rio Declaration 1992 effectively treated women's issues as a cross cutting theme and provisions were made in sections where women's issues needed to be mainstreamed (e.g. deforestation, education). It also listed areas requiring urgent action to avert ongoing environmental degradation that affects lives of women and children—draught, desertification and deforestation (Global

Assembly, 1992). The Rio Declaration, however, ensured the need of women's integration in addressing environmental issues for sustainable development. Mellor (1996) argues that since women are more involved in the work of caring and nurturing, so they are more attached to the environment than men. The activities like nurturing and caring is regarded as feminine work whereas mining and industrialization is regarded as masculine work where there is no place for women (Ballad and Banks, 2003). This has not only led to the gender division of labour in the industrialization process and its related activities but also negative impact on environment in terms of sustainable environment and development. This section of the paper examines the theoretical underpinnings of women, environment and livelihood issues of tribal women in one of the eastern states of India i.e. Orissa. It deals with the interrelationships and different dynamics that influence the relationship between environment and livelihood options of tribal women in Orissa.

Women and Environment Linkages

People live with all its specific environmental, human, social, economic and cultural characteristics in a livelihood pursuit (Scoones, 1998). The physical environment varies from place to place all over the world, thus the livelihood attached to it. The very set-up of the society also constitutes of diverse groups of people belonging to different age, sex, caste, class, religion and cultural backgrounds. Furthermore, gender relations determine the type of work that women carry out and their responsibilities and rights (NEDA, 1997: 1). The division of roles based on sex is almost everywhere in the world determined in relation to the traditional economic structures (Sen, et al., 1982). The risky tasks and works that need physical strength has been taken up by men whereas those works which requires sustained effort and is important for survival has been assigned to women which is further strengthened by the taboos and beliefs (*ibid.*). Again environment is experienced differently by men and women as a consequence of different context in which they operate i.e. social, cultural, economic and political environment. 'Val Brown argues that women's uses of the environment are adequately different from those of men as to constitute a distinctive "habitat", in the ecological sense' (Brown, 1992 as cited in Buckingham, 2000: 102). This is correct particularly where roles are assigned on the basis sex-division of labour and it is true for those women whose dependence on forests is much greater (Agarwal and Narain, 1985a as cited in Fernandes and Menon, 1987). 'This difference in experience and gender division of labour constitutes a significant element of what has been referred as *place identity*—that element of person's sense of self which derives from their experience of, and attitudes towards, the physical environment' (Buckingham, 2000: 102). Women play an important role in the management of the environment at the local level. Davidson and Myers (1992) argued that women are responsible for the primary environmental care. But it is place and space contextual.

The fact that women are closely linked to the environment has been recognized and has drawn the attention of the development agencies at the international, national and local levels to address it. Several steps have been taken by these agencies to focus on the environmental issues and gender to promote sustainable development through sustainable livelihood approach for the people, communities and countries. There are various diverse issues involved in the whole process of dealing with environmental issues, instead of dealing it in a purely sectoral way. The social, economic, physical, political and cultural aspects seem to be inseparable, so there is need to look at these issues in an integrated way. In such a context, an attempt has been made to study the inter-relationship between environment and livelihood.

Conceptualizing Livelihoods and Environment Linkages

Environmental resources provide life and livelihood to the poor people in the society. In the case of tribals, majority of them depend on the environment for their living and livelihood. Thus, any form of degradation of the environment by the rich or poor directly affects the people who are dependent on environment for their livelihood (Parikh, 2002). In the different levels of societies like traditional and tribal, urban and rural, agrarian and industrialized, different categories of people relate to the environment differently. The basic necessities of life such as water, fuel, food material and other resources are often provided by the environment. The contribution of environment in the lives of the tribal community plays a major role in providing them their livelihood. Therefore, any sort of threat to the environment is directly or indirectly threatening for the people dependent on it for their existence and livelihood.

The livelihood and environment linkages are essentially context specific and vary according to both the geographical location and economic, social and cultural characteristics of individuals, households and social groups. The people residing in the tribal belts have limited assets and have greater dependence on common resources for their livelihoods, so any form of environmental degradation impacts them much more than the others (UNDP, 2002). In addition, national and international political processes also affect people's environments and livelihoods. A country's development model as well as its politics and natural resource management, culture of production and consumption pattern influences the nature and dynamics of livelihoods and environment (*ibid.*).

Livelihood, Environment and Tribal Women in Orissa

In India majority of the tribal population resides in the rural pockets of the country. They depend directly or indirectly on the natural resources and ecological services for their existence and livelihoods. An estimate of 60 per cent of the country's workforce depends on agriculture, fisheries, forests and other natural resources for their livelihoods and most of them belong to the poor community (GOI, 2001). The tribal parts of the country present an atypical situation of being resource rich-regions but are characterized by underdevelopment and famine. The development in these areas has witnessed repeated droughts leading to famines, steady impoverishment of the bulk population, extreme hunger and starvation. Several factors have contributed to this. 'First, all lands for which individuals did not possess legal rights was declared as government or public property, including forests and hills. This led to the curtailment of tribal customary rights and drastic reduction in their area of operation. The system of shifting and fallowing in agriculture was affected, with the fallow period declining from 18-30 years to just three years, thus depleting the soil. Many of the traditional practices of sustainable agriculture in the forests and on hill slopes had to be abandoned due to deforestation and as such survival became the main concern' (UNDP, 2002: 1). A study by Bina Agarwal (1995), shows that the gender, environment and poverty vulnerabilities including livelihood issues often overlap in many regions such as Bihar (erstwhile), Uttar Pradesh, Rajasthan, West Bengal, Orissa and Madhya Pradesh (erstwhile). It indicates that the intensity of the problem varies from the region to region and the livelihood problems experienced by women of the different places differently.

In this section, an attempt has been made to understand and explain the relationship between tribal women, livelihood and environment in Orissa. Orissa has a large number of schedule tribe populations (23%), and these groups have strong links with the forest. The Schedule Tribes of Orissa are by tradition closely linked with and more reliant on the forest (Poffenberger, 1996), although they had little involvement in the decision-making in terms of use, control and access to forest and its resources due to their lower social standard (Sundar, 1997). However, the tribal economy also seems to be based on a division of labour based on a division of labour along gender lines. In Orissa, tribal women do majority of the work related to food procurement in the form of collection of minor forest produce, carrying water from the streams and rivers etc. They even take care of the children and perform all the household activities. Modern forestry matters, such as timber, are seen as a male issue (Shiva, 1997). This has been shown through labour division within families and in the set up of the government department like Forest Department where hardly any women have been employed. It is here where forest products focus on timber. Other products, such as mushrooms, fruits, roots, and leaves are mostly dealt with by women and are regarded as 'minor' even though they are a vital source of livelihood for many families in India (Malhotra, 1997) and the collection of these are considered women's work (Shiva, 1997). Though it may not be true for all the women in Orissa, the tribal women are considered as the primary caretakers of environment because of their relation and linkage with the environment for livelihood be it forests, rivers or any other natural resources. Culturally tribal women are the real bread earner of the household. They are responsible for the management of the family income by collecting and selling forest products. In tribal societies women have received higher status due to provision of cash income from the sale of non-wood forest products (Yadama et al., 1997) that makes them more equal to the men compared to women belonging to other castes and communities. However, the importance of forest products is different between the sexes as well, due to variations in responsibility in the home and society (Chatterjee and Roy, 1995).

The tribal women are the key players in acting as carers for their surrounding environment. Because of their relationship with environment for survival and living, the tribal community had developed a culture related to the conservation of natural resources. 'The destruction of forests, which is their life system, has resulted in their impoverishment, indebtedness and in many cases land alienation and even bondage. While the tribals were impoverished in general, it was found that women faced more adverse affects as their links with forests are much closer than that of men' (Fernandes and Menon, 1987: 1). It has been pointed out that women are victimized by environmental degradation since their working days has been immensely lengthened by scarcity of water, fuel and fodder, and their traditional skills and occupations have been unfavourably affected by new technologies in agriculture, artisanal work, industries and market, while new opportunities have not sufficiently developed (Agarwal and Narain, 1985). In order to challenge these broader processes of environmental deterioration, women have both practical skills and mental resources which need to be mobilized. It is also necessary to work out the relationship between environment and cultural crisis caused by development paradigms which is neo-colonial, capitalist and patriarchal in nature and also violently hamper the base for human survival and their livelihoods that is destructive in both nature and culture (Dietrich, 1999).

The addressal of environmental degradation will not only provide sustainable livelihood but also reaffirm the idea of sustainable environment. Thus the issues of environment and livelihood often require cross-sectoral approaches and which very much goes beyond agricultural issues. The initiatives must focus on the diversity of livelihoods sources and analytically address conditions that bar the ability of the tribals to overcome poverty (OECD, 2001). The existing relationship between tribal women and environment has a major problem of survival and identity (Dietrich, 1999). The environmental crisis, whether that of forest or any other resource, gains relevance only when related to the people who are affected by it (Viegas and Menon, 1985: 57 as cited in Fernandes and Menon, 1987). 'Consequently, industrial development which is necessarily geared to resource utilization rather than resource conservation' (Simmons, 1974 as cited in Fernandes and Menon, 1987: 4) has led to the massive environmental degradation which has direct impact on the livelihood of the people who are depending on it more i.e. the tribal people.

PART-2

Tribal Women and Deforestation: The Case of Industrialization in Orissa and the Dilemma of Development

The previous part has seen the embedded relationship between tribal women, environment and livelihood with special reference to Orissa, one of the eastern states in India. Though the tribal women occupy a better position in their communities compared to other women, they seem to be the worst victims compared to their men in the course of development and industrialization. This part explores the present industrialization process which is said to be violating all environmental regulations of the government by its own institutional set ups. While examining the industrialization process, the chapter encounters the dilemmas of tribal development where the livelihood of the tribal people and women in particular are in trouble.

The process of industrialization in India has been a conflicting issue for the development of the nation since Independence. The imagination of nationhood was associated closely with the

Gandhian vision of the rural economy as the foundation for development while this contradicted with the commencement of Nehruvian ideals of 'progress' through aggressive modern development strategies (Chatterji, 2001). Gandhian views stressed the usage of appropriate and local technologies to provide employment and livelihood to India's agrarian population. Mahatma Gandhi promised a liberation that included economic and social security, premised on a sacred commitment that the diversity of cultural, social and spiritual traditions of India be preserved while feudal, colonial and postcolonial oppressions be addressed (Baviskar, 1995). Later on the vision's of Gandhi was put aside by the ideas of Jawaharlal Nehru and others who opted for development through large scale industrialization, urbanization and modernization. This approach, designed to eradicate poverty and debt, paradoxically targeted elite and urban sections of India with residual impact on rural and tribal populations (Chatterji, 2001). This aggressive mode of development in the form of industrialization seems to have hampered the progress on the poor people in several states of India. Orissa too has been affected as it is one of the natural resource rich states in India that is till today victimized in terms of modern development interventions.

Before exploring the implications of industrialization on the lives of the tribal communities and on women in particular, there is the need to get an overview of the state of Orissa. 'The state of Orissa hosts more than 11.4 per cent of the total tribal population in India and it accounts for 62 communities. Around 44.7 per cent of the total area in the State is declared as the Schedule Area[2] under the Fifth Schedule of the Constitution of India. The forest area in the State is 9,963.30 sq km. constituting 38.48 per cent of the total area. Four districts namely Sundargarh, Mayurbhanj, Koraput and Rayagada out of the total of thirty districts in Orissa are predominantly inhabited by the tribals. The overall economy of Orissa remains backward though it is rich in natural resource potential. More than 64 per cent of the people live below poverty line. Per capita income is far below the all-India average, one of the lowest in the country. The total cultivated area in Orissa is around 40 per cent and the net irrigated area as per cent of net sown area is 19 only. The industrial dimension in Orissa is still bleak, as the production is 1.6 per cent. The manufacturing sector employs only 7 per cent of the workforce and less than 35 per cent of the workers are in non-household industrial sector. Still there are large untapped resources, widespread sharecropping, unemployment, feudal strongholds, indebtedness, etc., which cumulatively contribute to the underdevelopment of the state' (Pathy, 2003: 1).

Industrialization and Endangered Livelihood of the Tribal Women: Case Studies from Orissa

Although very little industrialization has taken place in Orissa, this process seems to have violated all the environmental regulations of the government and have made them more vulnerable by taking away their resource base for livelihood and existence. 'The optimistic assumption that increased welfare due to industrial growth would automatically percolate to the poor has not been borne out of experience' (Baviskar, 1995: 25). Some case studies from Orissa depict the brutal face of industrialization. The Independent magazine (2006) refers to one of the case studies of industrialization in Orissa. It quotes that India's economic success is a modern miracle. But the dark side of the boom has been its tragic cost to the subcontinent's most vulnerable people (means women and tribals). It relates an event that took place on the 2nd of January 2006 in the quiet morning rituals of Kalinganagar, a village in the Jajpur district of Orissa. 'A severe massacre of tribal people took place by armed police people in order to obtain their land for industrialization purposes. Tata, a major Indian company, wants to build another very big steel industry compared to the already existing three steel plants in its surrounding areas by the Jindal, Visha and Bhusan. The villagers of Kalinganagar belonged to the indigenous Ho tribe. In this massacre 12 people were slaughtered and 31 were injured.' It is a reflection of the Government of Orissa's commitment to industrialization in the state at the cost of human rights violation by its own institutional set up i.e. the Orissa Police.

Another case is of the Kashipur block of the Rayagada district where three tribals died because of the police firing in

Maikancha village in the year 1999 and few deaths took place in the year 2002 due to poverty but few suggests that some one killed them (Das, 2003). 'On 1 December 2004, the Orissa police attacked and critically injured 16 tribals in the same Kashipur block. Many, disproportionately women, were arrested. They were targeted for protesting against the proposed aluminium plant site of Utkal Alumina International Limited (a joint enterprise of Aditya Birla Group, and ALCAN, a Canadian company) at Doraguda. The project is expected to displace and dispossess 20,000 people and impact rights of life and livelihoods across 82 villages' (Chatterji, 2005: 1).

In both the cases it seems that the Government of Orissa betrays its legal and ethical mandate by suppressing public dissension through police brutality (Chatterji, 2005). The mining industries are the focal industries today where large scale environmental degradation and violation of human rights take place. These abuses at times are conducted by the Orissa police. This indicates a tension between the government policy that emphasizes the importance of people's environmental and livelihood rights and local policing procedures.

Present Scenario

The way the present paradigm of development projects in India, are taking place based on the idea of economic growth. The consequences for people and environments in these countries are unprecedented in scale and speed as a scramble for resources takes place to feed the demands of industry in both; while environmental governance has failed to keep pace with the speed of change (Bauer, 2005). In India the exponential rise in the demand for timber, minerals and water which dictates the speeds of change in the landscape has taken place since about 1980 and has been hastened in the by Indian 'economic reform' since 1991.

The greatest physical impact of industrialization/mining on people and landscape in India is being exerted in the mountainous parts of Orissa, Jharkhand and Chhattisgarh, coincidently amongst the poorest states of India. This is because they possess untapped sources of iron, bauxite and water reserves, all vital constituents of the spiralling Western demand for minerals and metals (Padel and Das, 2006). A high proportion of the rural population of these heavily forested regions are tribal groups that are highly vulnerable to the processes of modernization and land alienation that accompany the huge pressures for mineral extraction and dam building that illustrates the economic globalization that is accelerating and transforming the region.

Forests in India have shrunk to the alarmingly low amount of 21 per cent of total ground covers forests, while Orissa which had the proud privilege of impregnable forest has gone down to 14 per cent (Patnaik, 2003). About 4.26 per cent of people belonging to the tribal groups practice shifting cultivation which is locally called 'podu' (Pathy, 2003). Often the tribals residing in these areas are blamed by the forest officials for forest extinction because of their shifting cultivation activity to earn their livelihood. But the tribals too have an answer for the regeneration of forests. They leave stumps of trees intact to regrow in the forest areas used for shifting cultivation. It is a common notion that the poor people are blamed for destroying the environment, but in fact it is often rich, powerful and wealthy interests that are doing this (Arora, 1994). But in the name of development and industrialization other hills covered with dense forests have been blasted out for iron and bauxite without any consideration for the replantation of the area. Orissa in particular has been targeted as it is one of the poorest states in India with dense forest and rich mineral resources (*ibid.*).

Industrialization and its Impact

Land is considered as the life of people. 'Take his land away and you have taken away his life.' This old saying has proven to be true in the mineral and resourceful districts of Orissa. The tribals, who depend on water, forest and land for their livelihoods are on the verge of virtual extinction not only socio-culturally and psychologically, but also physically, economically and politically (Kujur, 2005). Land acquisition by mining industries affect the communities in a similar manner (or perhaps more severe in nature) than land acquisition for dams and other big projects. When mining industrialization takes place the immediate threat to the local communities who are mostly tribals is displacement and land alienation, which is a direct threat to their livelihood. Minerals are considered as national resources and subsoil rights over lands belong to the state and not to the individual owners (Mines, Minerals and People, 2003). 'The powers of eminent domain, which are enjoyed by the state while acquisition of lands for mining, positions local communities in totally helpless situations—mines and communities' (*ibid.*). According to Bebbington (1999: 2022), the poor are seen as the potential agents of change: '...people's assets are not only a source of sustenance and meaning—they are also a source of power'. It seems clear from such initiatives of the Indian state that the tribal communities are not only losing their livelihood but also the source of power is strategically taken away from them in the name of industrial development. It is estimated that some 3 to 5 million people have been displaced since 1950 in Orissa on account of various development projects. More than 50 per cent are tribals and the expected displacement in coming decades also is expected to affect tribals disproportionately (Saxena, 2001).

Tribals and particularly tribal women who depend on forest products for their daily livelihood are often treated badly in the hands of the forest guards and other officials. Cases like sexual assault and harassment has been recorded. "In discussion recently with the National Women's Commission it was suggested that a record be compiled of the number of cases against tribal women by the forest department or other authorities for trespassing and using forest products. In 2004 it was found that in Orissa there were as many as 11,424 cases by the Forest department on tribals, men and women involving 'theft' of forest produce valued at less than 100 rupees" (Karat, 2006: 4). It is shocking realisation that gives a glimpse of a horrible situation that the tribals' face. The control over minor forest produce is extremely important if the traditional livelihood of forest dwelling communities is to be protected (*ibid.*).

Mining as the focal industry in Orissa has increased the exploitation of women's rights with regard to land and livelihoods. With regard to land, women have no legal rights over land and the natural resources. In some places where the tribals are educated, they have legal right over land but in most of the cases they do not have legal right due to their illiteracy. They do enjoy greater social status compared to other rural women with regard to control over the resources because of their active participation and decision-making with regard to land utilisation, agriculture and powers over cash flow in a tribal economy. 'However, both rural and tribal women are completely alienated from formal access and rights when mining industries come in. When villages are affected or displaced in order to make room for big industrial developments like mines, women have been forced out of their land based work and pushed into menial and marginalised forms of labour as maids and servants, as construction labourers or in prostitution, which are highly unorganised and socially humiliating' (Mines and Communities, 2003). For example, women as losers of land and as displaced persons from the coal mining areas of Talcher (Orissa) faced serious problems affecting their lives, as their link to livelihood, economic and social status, health and security all depended on their access to land and forests (*ibid.*). The economic loss of tribal women and their families who never received compensation or rehabilitation is far more severe. In such cases these people migrate to neighbouring states in search of other forms of livelihood. One example concerns tribals from Orissa who migrated to Andhra Pradesh, a neighbouring state, where they occupied lands high on the hills or encroached on forest lands at the risk of being 'criminals' in the eyes of law in order to survive.[3]

This chapter clearly depicts the horrors of industrialisation which have influenced the lives of the tribal communities and more severely to the tribal women. This does not mean that industrialisation is not essential in Orissa for development. It can be very fruitful if it is planned and implemented properly taking into consideration the needs and interests of the tribal communities or any other people who are going to be affected by the set up of the industries. The next part will focus on the environmental governance issues and will try to locate the role of state and its policies related to environmental management affects the livelihoods of the tribal women.

PART-3

Implications of Industrialization on Tribal Women and their Livelihood in Orissa: The Role of State and Public Policy in this Regard

It is very clear from the previous chapters that men and women are involved in different spatial and qualitative areas of activities: private, public and intermediate; paid and unpaid work. It has been found that the men are more involved in public and paid work sector which elevates the unpaid work or unorganized sector in which women are usually involved. This imbalance has implications on women along with other vulnerable sections of the society. The issues of livelihood related to environment too has the same impact and (Buckingham, 2000: 103), points out that the specific environmental concerns relates to 'those with undervalued and unrecognized contributions to society who are more likely, because of a lack of financial wealth, to suffer environmental problems which they cannot easily escape. These people are least able to do anything about these environmental problems because of their lack of a recognized public role, which effectively confers on them second-class citizenship.' Accordingly, this chapter deals with the implications of the industrialization process, which is one of the environmental and developmental concerns, on the livelihood of the tribals in Orissa where women are the worst victims being the bread earner of the tribal family. It will further reflect on the role of state and public policies as they relate to environmental management which, in turn, impacts the livelihoods of the tribal women.

Role of Industries and Policies in Environmental Management

The debates about the role of industries in environmental management and the ways in which policy influences the livelihood of people has been taken up by environmental social science researchers. This research has played important role in bringing forward the discussion around the policies and the livelihood of people. 'Debate about the effects of economic growth on the environment has been long polarized. On the one hand, there are some who argue that growth brings prosperity, changes in the structure of industry, new technology and social expectations that enable societies to invest in improving their natural environment' (Berkhout, et al., 2003: 19). On the other hand, critics argue that the rich countries use more natural resources and contribute to large scale environmental impacts like climate change and destruction of biodiversity (*ibid.*). They argue that there are limits to the usage of environmental services beyond which they will be endangered and the overall welfare of the people will be threatened (de Groot, 1992, Daly, 1994). In Orissa too the industrialization process was initiated with a motive to enhance economic growth and employment opportunities but at the cost of the welfare of the poor people. As such, it was assumed that industrial growth would be a positive process which would not destroy natural resources and environmental conditions.

Moreover, in Orissa policy making links directly was drawn between poverty and environmental degradation, was also assumed that being that poor rural people would have no choice but to 'mine' environmental resources to live their life (Durning, 1989). This conservative policy approach has been criticized for failing to recognize that numerous cases exist where local people effectively conserved and nourished the environment despite immense material poverty (Fairhead and Leach, 1998, Scoones; 2001). The study by Fernandez and Menon (1987) found out that there is a growth of forests while it is under community management in Orissa. However, most of the states in India have predominantly agrarian population dependent on the land, forests and other natural resources for its sustenance and livelihood, socially, culturally and economically. This tends to make it difficult in drawing straightforward correlation and link between poverty and environmental degradation. Moreover, people's behaviour and the associated environmental impacts were shown to be influenced by a wide range of bridging factors like knowledge and values, social difference and dynamics, and institutional and policy arrangements governing both access to, decision-making and control over resources, to the terms on which technologies are employed (Leach et al., 1999). In Orissa rural and tribal women are the primary actors in agriculture, collection of forest produce, in livestock management apart from caring their families. It has been accepted as a global fact that women, rural and tribal, have a very intimate and symbiotic relationship with the environment around them as they are inseparably linked to the natural resources. It is important that this link between women and environment is well understood when development models and shifts are logically reasoned by governments and societies to improve socio-economic situations.[4] However, in reality women have often been alienated from conceptions of development and their close association with their environment ignored. But, 'policies cannot be premised on an assumption of a generalized affinity between women and nature, or simplistic observations of 'what women do'—there are serious risks of simply adding environment to the already existing list of women's caring roles, instrumentalizing women as a source of cheap or unrewarded labour' (Leach, et al., 1995: 7).

Role of State and Policies in Orissa

In Orissa, development became more serious after the 1950s with about 80,000 families affected and displaced in order to make way for development projects. State-driven, large scale industrial development was the primary focus and companies such as TATA, NALCO, HAL, JINDAL, HINDALCO, RIO TINTO, ALCAN and POSCO were established here. As mentioned in earlier chapters, huge development projects were initiated in Orissa which boosted in the early 1950's. Huge dams were constructed and number of industries was launched to boost the economic growth of the state. Huge displacement of people along with acquisition of land took place by such developmental processes. Only in the year 1973, Orissa came up with its first resettlement policy and in 1977 approved land for land as the basis for rehabilitation. The mentioned statistics given in the introduction part of displacement (81,176 people) and land acquisition (14,82,626 acres) gives the picture that for long the state had its eyes blind folded towards the development of project affected people and their relationship with the environment. "Nature has no power or will, and does not exact revenge when we transgress her laws. In as far as humans, socially organizing themselves, do degrade the natural environment it is our social 'fault', and not outside our control; though of course it is the powerless who invariably suffer the full effect of environmental degradation and are politically least able to alter the course of events" (Cole, 1999: 142).

But the project-affected communities cannot withstand this situation for long. For several years, local communities in Rayagada, Kalahandi and recently in Kalinganagar have been protesting bauxite and iron-ore mining by a group of industries. They have been condemning the breach of constitutional provisions barring sale or lease of tribal lands without their consent. People opposed the destruction of their ecosystems, futures, the destruction of forests, agricultural lands, mountains, perennial water-streams, the water retention capacity of mountains, all of which is integral to their lives and livelihood. 'Orissa is a tragic affidavit of the intersections of irresponsible globalization, State complicity in defiling human rights, and police participation in fostering social violence' (Chatterji, 2003). Mining initiatives in these areas in Orissa contradicts the United Nation's Declaration to the Right to Development, which mandates 'free and meaningful participation' in development. 'It violates the right to life and livelihood guaranteed by Article 21 of the Indian Constitution, and contravenes the directives of the National Forest Policy, 1988, which legitimates the traditional claims of forest dependent communities to public resources' (*ibid.*). The state did not take any initiative to involve the residing communities while taking decision to start industries in these areas. Here the state seems to exercise its power to take over their land and forests leaving them in a vulnerable condition. It also keeps its eyes closed to the 'Provisions of the Panchayats (Extension to the Scheduled Areas) Act' of 1996. 'PESA' seeks to provide significant protection to the tribals in the scheduled areas against any arbitrary, discretionary and motivated action by the state relating to land acquisition and resettlement and rehabilitation (R and R) package for the project affected people. This applies to the tribal areas of eight states in India which are guided by the Schedule V and VI of the Indian Constitution and Orissa is one amongst them. It further mandates that "gram sabha or the panchayats at the appropriate level shall be consulted before making the acquisition of land into the scheduled areas for development projects and before resettling or rehabilitating persons affected by such projects in the scheduled areas" (Bandhyopadhyay, 2004).

In the name of economic growth of the state, maldevelopment seems to have taken place, endangering people who depend on natural resources for subsistence. This is of particular concern in the context of growing liberalization and corporate globalization prioritized by the State in trade, industry, tourism and agriculture, and through the privatization of public resources and infrastructure (Chatterji, 2005). 'The State often charges poor rural communities with the primary responsibility for ecological degradation, while plans for allaying rural poverty emphasize capital and resource intensive strategies as devised by the National Forestry Action Plans prepared by the Ministry of Environment and Forests in 1999. The Revised Forest Strategy of the World Bank, approved in October 2002, is another example of centralized policies that alienate the poor by privileging 'free' market activity through endorsing the unchecked involvement of the private sector in development processes' (*ibid.*).

Moreover, regarding the issue of land acquisition, there are complex procedures for doing it and for obtaining mining leases for exploration, prospecting and extraction, in reality, mining projects are sanctioned more with the view of encouraging and catering to the interests of the mining industries rather than to protect the interests of the local communities and the rest of the natural resources which exist in a mining area apart from the minerals, as reflected in the New Mineral Policy of 1993 (Mines, Minerals and People, 2003). The Land Acquisition Act of India is also threatening and gives over-riding powers to the state to encroach onto people's lands for any 'public purpose' including mining. In most mining situations, communities are not even aware that their land has been offered on lease to big industrialist for mining or other big projects. It is only at the time of eviction when the bull-dozers are brought in, many a times with police presence that the communities realize that their land has been leased out by the state without their knowledge. The 'public purpose' of a mining activity from a gender perspective and especially with regard to what the women in the communities stand to 'publicly' gain, is highly uncertain (Mines, Minerals and People, 2003).

According to Bandhyopadhyay (2004), a responsive and a humane government would have taken this unrest seriously, would have initiated dialogue and interface meeting initiatives to build confidence between the project affected people and the government. It was only then that the 'bitter' pill could have been accepted by the former. 'Instead of doing that, the government had been acting as the managing agent of the corporate giants who are setting up bauxite mines and alumina plants' (*ibid.*). This total lack of trust and faith in the state by the people has made the situation worse and a complex problem has developed. In these respects, contemporary perspectives on sustainable livelihoods in the south (Scoones, 1998 as cited in Berkhout, 2003) converge with northern work on sustainable lifestyles and consumption in emphasizing the broader institutional and policy processes that shape individuals' and group behaviour, and the need to address strategies for change towards these.

However, the deadly clash that took place in Kalinganagar, Orissa, on 2nd January, 2006, led the state government to frame their Rehabilitation and Resettlement policy which includes the interests of the affected communities. This draft seems to be more sensitive towards women as they have provisions for women which make them non-dependent on others. It has clearly stated that the benefits of the policy will not go to the son only but will go to daughters, divorcees, widows, etc. It also has the provision of trainings which will not be provided to men only but women too will be included in it for employment in the industries. With many other provisions in it that favours the project affected communities, this document seems to be a considerate document. The two important clauses of this draft include—Clause 5(A)(b) which says—"For rehabilitation and displaced families, R&R packages would be implemented as per prevailing guidelines and practices of government of Orissa." Clause 5 (G)(a) states—"In the matter of employment, preference will be given to the people of Orissa subject to their possessing the necessary qualifications; [concerned industry] will make every effort to improve their skill levels, if necessary through specialized training. For this purpose state government will nominate a nodal agency to coordinate with the [concerned industry]." It is hoped that this draft policy, 2006 will act as tough gatekeeper for the industries who are going to come and utilize the rich mineral and forest resources of Orissa.

Conclusion

Forests and forest resources, primarily minor forest products (MFP) play an important role in the survival of tribal households in Orissa and elsewhere in India, because of the importance of forests in their social, cultural and economic survival (Tewari, 1989). Studies conducted in the tribal regions of north-east India also indicate high rates of dependence on forests to meet fuel wood needs (Maikhuri, 1991). Recognizing this importance of forest resources to tribal communities, it is necessary to draft tribal a sensitive forest policies to sustain their livelihood. Others have noted the important role of forest resources not only in meeting the subsistence needs but also in poverty alleviation of tribals (FAO, 1995). It is again interesting to note that tribal economy revolves around tribal women as they are the primary players in the collection, processing, and marketing of forest resources and bulk of the agricultural produce, including food and fuel-related forest products. A study conducted by Fernandes and Menon (1987) on Orissa, reported that women generally walk 3 to 4 hours into the forests and work 15 hours per day, whereas men work 11 hours a day. The studies all indicated that women spend more time and labour in forest-related activities, and depend on forests not only to meet subsistence needs but also for income. Interestingly, the considerable role played by women in ensuring food security, and in the provision of cash income from the sale of different forest products gives women a higher status in tribal societies (Sarkar, 1994). This indicates that the relationship of tribal women with forest and natural environment not only provides livelihood but also empowers them and put them in a higher social and economic status.

The industrialization process in Orissa is a threat not only to the environment and forests but also a threat to the livelihood of the tribal women. It makes tribal women more vulnerable than the men due to the fact that tribal women are more dependent on forest for their household livelihood than tribal men. Again industrialization and "mining is an exceptionally masculine activity in terms of its workforce, its cultures of production" argues Robinson (1996: 137). There is no place for women to work in such industrial and mining projects. It indicates that these projects have taken away forests and land which was the source of livelihood for tribal women and their families. In return, the tribal women are put into a situation of vulnerability. Apart from the economic impact of this vulnerability, its social impact is more far-reaching. It has changed the social and economic status of tribal women in their society through the process of economic disempowerment. In this way the industrialization and mining projects has not only changed the ecology and livelihood of the tribal women but also contributed for the development of a gendered unequal society by putting women in a lower social and economic status in Orissa. Lastly, it can be said that industrialization and mining in Orissa is another mode of marginalization for tribal women.

NOTES

1. Although the Constitution of India does not define Scheduled Tribes as such but Article 366(25) of the Constitution of India refers to scheduled tribes as those communities who are scheduled in accordance with Article 342 of the Constitution. According to Article 342 of the Constitution, the Scheduled Tribes are the tribes or tribal communities or part of or groups within these tribes and tribal communities which have been declared as such by the President through a public notification. Scheduled Tribes are spread across the country mainly in forest and hilly regions. They are also known as the indigenous people, original inhabitants (*adivasi*), forest dwellers (*vanavasi*) and 'tribals' constructed by the colonial British rulers.
2. The term 'Scheduled Areas' has been defined in the Indian Constitution as "such areas as the President may by order declare to be Scheduled Areas". Paragraph 6 of the Fifth Schedule of the Constitution prescribes following procedure for scheduling, rescheduling and alteration of Scheduled Areas. The criteria followed for declaring an area as Scheduled Area are preponderance of tribal population;

compactness and reasonable size of the area; underdeveloped nature of the area; and marked disparity in economic standard of the people. These criteria are not spelt out in the Constitution of India but have become well established. They embody principles followed in declaring 'Excluded' and 'Partially-Excluded Areas' under the Government of India Act 1935, Schedule 'B' of recommendations of the Excluded and Partially Excluded Areas Sub Committee of Constituent Assembly and the Scheduled Areas and Scheduled Tribes Commission 1961 (Ministry of Tribal Affairs, Government of India, New Delhi).

3. http://www.womenenvironment.org/mining2.asp visited on 10th July, 2006.
4. http://www.womenenvironment.org/mining2.asp accessed on 10th July, 2006.

REFERENCES

Agarwal, B. (1995). Gender, Environment and Poverty Interlinks in Rural India: Regional Variations and Temporal Shifts, 1971-1991, Discussion Paper prepared for United Nations Research Institute for Social Development.

Agarwal, N. and Narain, S. (1985). *The State of India's Environment: A Second Citizen's Report,* Centre for Science and Environment, New Delhi.

Arora, Dolly (1994). "From State Regulation to People's Participation: Case of Forest Management in India." *Economic and Political Weekly,* March.

Ballard, C. and Banks, G. (2003). "Resource Wars: The Anthropology of Mining." *Annual Review Anthropology,* 32.

Bandhyopadhyay, D. (2004). "Rayagada Story Retold, Destitutes of Development." *Economic and Political Weekly*, 31 January.

Barraclough, Solon L. and Krishna, B. Ghimire (1995). *Forests and Livelihoods: The Social Dynamics of Deforestation in Developing Countries,* MacMillan Press, New York.

Bauer, J. (2005). *Forging Environmental Values; Contested Environments in China, Japan, India and the United States,* M.E. Sharpe.

Baviskar, A. (1995). *In the Belly of River; Tribal Conflicts over Development in the Narmada Valley*, Oxford University Press, New Delhi.

Bebbington, A. (1999). "Capitals and Capabilities: A Framework for Analysing Peasant Viability, Rural Livelihoods and Poverty." *World Development*, 27, 12.

Berkhout, F., Leach, M. and Scoones, I. (2003). *Negotiating Environmental Change, New Perspectives for Social Science,* Edward Elgar Publishing Limited, UK.

Buckingham-Hatfield, S. (2003). *Gender and Environment,* Routledge, London.

Chatterjee, M. and Roy, S.B. (1995). *Reflections on Gender Issues in Joint Forest Management,* Inter-India Publications, New Delhi.

Chatterji, A.P. (2001). "A Critique of Forest Governance Eastern India." *International Journal of Economic Development,* Vol. 3, No. 2.

Chatterji, A. (2005). "State Tyranny in Orissa." *Sikh Spectrum. Com. Quarterly,* Issue No. 19.

Chopra, K., Kadekodi, G.K. and Murty, M.N. (1990). *Participatory Development: People and Common Property Resources.* Sage Publications, New Delhi.

CSE (1985). *State of India's Environment—1985,* Centre for Science and Environment, New Delhi.

Cole, K. (1999). *Economy Environment Development Knowledge,* Routledge, London.

Daly, H. (1994). "Operationalising Sustainable Development by Investing in Natural Capital." Jansson et al., *Investing in Natural Capital: The Ecological Economics Approach to Sustainability,* Island Press, Washington, D.C.

Das, V. (2003). "Kashipur: Politics of Underdevelopment." *Economic and Political Weekly Discussion,* 4 January.

Davidson, J. and Myers, D. with Chakraborty, M. (1992). *No Time to Waste: Poverty and the Global Environment,* Oxfam, Oxford.

de Groot, R.S. (1992). *Functions of Nature,* Amsterdam, Wolters-Noordhoff.

Dietrich, G. (1999). "Women, Ecology and Culture." Menon, N. (ed.) *Gender and Politics in India,* OUP, New Delhi.

Durning, A. (1989). "Poverty and Environment: Revisiting and Downward Spiral." Worldwatch Paper 92, Worldwatch Institute, Washington D.C.

Fairhead, J. and Leach, M. (1998). *Reframing Deforestation. Global Analysis and Local Realities: Studies in West Africa,* Routledge, London.

FAO, 1995. Report of the International Expert Consultation on Non-Wood Forest Products. Food and Agriculture Organization of the United Nations, Rome.

Padel, F. and Das, S. (2006). "Adivasis Shining: Standing up to the Face of an Invasion of Aluminium Companies in Orissa." Nayak, B.S. (ed.)

2006, *Nationalizing Crisis: The Political Economy of Public Policy in Contemporary India*, Atlantic Publishers, New Delhi.

Fernandes, W. and Menon, G. (1987). *Tribal Women and Forest Economy: Deforestation Exploitation and Status Change*. Indian Social Institute, New Delhi.

Global Assembly (1992) of Women and the Environment 'Partners in Life', Miami 1991, Summary for UNCED: PreCom IV Decisions, Follow-up, Issue-3.

GOI (2001). *Census of India: 2001*, Government of India, New Delhi.

Guha, R. (1983). "Forestry in British and Post British India: A Historical Analysis." *Economic and Political Weekly*, 18 (44, 45, 46).

Haeuber, R. (1993). "Development and Deforestation: Indian Forestry in Perspective." *The Journal of Developing Areas*, 27.

Jena, M. (2006). "Orissa: Draft Resettlement and Rehabilitation Policy, 2006." *Economic and Political Weekly*, 4th February.

Kalam, A.P.J. (1998). *India 2020: A Vision for the New Millennium*, Penguin, India.

Karat, B. (2006). "Towards a Campaign for the Proposed Amendments in the Tribal Rights Bill." *People's Democracy*, Vol. 30, No. 23.

Kujur, J.M. (2005). "'Development' not for Tribes, Mines and Communities." in http://www.minesandcommunities.org/Country/india17.htm accessed on 15-08-2006.

Mines, Minerals and People (2003). Land, Forests, Displacement and Livelihoods: Protecting Women's Rights in Mining, Background paper prepared for the Indian Women and Mining Seminar, Delhi.

Leach, M., Mearns, R., Scoones, I. (1999). "Environmental Entitlements: Dynamics and Institutions in Community-based Natural Resource Management." *World Development*, 27 (2).

Leach, M., Joekes, S., and Green, C. (1995). "Gender Relations and Environmental Change." *IDS Bulletin*, Vol. 26 (1), Brighton, IDS.

Maikhuri, R.K. (1991). "Fuelwood Consumption Pattern of Different Tribal Communities Living in Arunachal Pradesh in North East India." *Bioresource Technology*-35.

Malhotra, K.C. (1997). "Emerging Patterns of Forest Management in India: Problems and Opportunities." Sengnjita, S. (ed.), *Studies in Anthropology*, Inter-India Publications, New Delhi.

Mellor, M. (1996). "Sustainability: A feminist approach." Buckingham-Hatfield, S. and Evans, B. (eds.), *Environmental Planning and Sustainability*, John Wiley: Chichester.

Menon, G. (1991). "Ecological Transitions and the Changing Context of Women's Work in Tribal India." *Purusartha*.

National Tribal Policy (NTP) (2005). *A Policy for the Scheduled Tribes of India, 2005, Ministry of Tribal Affairs*, Government of India, New Delhi.

NEDA (1997-2001). "Gender and Environment: A Delicate Balance between Profit and Loss, Women and Development." Working Paper-1, Ministry Foreign Affairs, Hague.

OECD, (2001). "Poverty-Environment-Gender Linkages." *DAC Journal*, Vol. 2, No. 4, OECD, Paris.

Parikh, J. (2002). "Poverty-Environment-Development Nexus." *International Journal of Global Environmental Issues*, Vol. 2, No. 3-4.

Pathak, A. (1994). *Contested Domains: The State, Peasants and Forests in Contemporary India*. Sage Publications, New Delhi.

Pathy, S. (2003). "Destitution, Deprivation and Tribal 'Development." *Economic and Political Weekly Perspectives*, 5 July.

Patnaik, G.B. (2003). "Whispers of the Wild." Seeland, K. and Schmithusen, F. (eds.), *Indigenous Knowledge, Forest Management and Forest Policy in South Asia,* DK Printwala, New Delhi.

Poffenberger, M. (1996). "Valuing the Forest." Poffenberger, M. and McGean, B. (eds.), *Village Voices, Forest Choices—Joint Forest Management in India,* Oxford University Press, New Delhi.

Sarkar, S. (1994). "Status of Tribal Women in Three Socio-cultural Dimensions." *Man in India,* 74(1).

Saxena, N.C. (2001). "Empowerment of Tribals Through Sustainable Natural Resource Management in Western Orissa." Report for IFAD/ DFID.

Scoones, I. (1998). "Sustainable Rural Livelihoods: A framework for Analysis." Working Paper-72, Institute for Development Studies, Brighton.

Scoones, I. (ed.) (2001). *Dynamics and Diversity: Soil Fertility and Farming Livelihoods in Africa,* Earthscan, London.

Sen, J., Saha, N. and Mukherjee, S.K. (1982). "Role of Women in Tribal Economy." Singh, K.S. (ed.), *Economies of the Tribes and Their Transformation,* Concept Publishing Company, New Delhi.

Shiva, V. (1997). *Staying Alive: Women, Ecology and Development,* Zed Books, London.

Sundar, N. (1997). *Subalterns and Sovereigns*. Oxford University Press, New Delhi.

Tewari, D.N. (1989). *Dependence of Tribals on Forests*. Gujarat Vidyapith, Ahmedabad.

Robinson, K.M. (1996). "Women, Mining and Development." Howitt, R., Connell, J., Hirsch, P., (ed.), 1996. *Resources, Nations and Indigenous Peoples: Case Studies from Australasia, Melanesia and Southeast Asia*. Oxford Univ. Press, Melbourne .

United Nations (1995). Beijing Declaration and Platform for Action, Adopted by the 4th World Conference on Women: Action for Equality, Development and Peace, Beijing, 15th September.

Wysham, D. (2001). *Behind the Shining: Aluminium's Dark Side*, Washington: IPS/SEEN/TNI (Unpublished, MS).

Yadama, G.N., Pragada, B.R. and Pragada, R.R. (1997). "Forest Dependent Survival Strategies of Tribal Women: Implications for Joint Forest Management in Andhra Pradesh, India." FAO Rap Publication, 24.

News Article

The Independent (2006). "India: The Terrible Price Paid for Economic Progress." *The Independent*, dated 11-03-06.

Collective Action for Sustainable Forestry: Institutional Dynamics in Community Management of Forest in Orissa

8

Satyapriya Rout

I. INTRODUCTION

The disappointing results of natural resource conservation policies in developing countries and the subsequent depletion of forest and other natural resources over the past few decades, have forced the scholars, development practitioners, and bureaucrats to shift their focus away from state-centred policies of conservation towards solutions at the local level. The resource degradation of the recent few decades has not only altered the world ecology and environmental scenario at the global level, but has challenged the livelihood security of thousands of indigenous population at the local level too. The intrinsic relationship that the communities share with their local resources and their inherent dependency upon those resources not undermining, it has been argued that policies regarding sustainability of the resource cannot be designed without integrating these communities in management strategy.

Acknowledging the role of local communities in halting the degradation of natural resources and promoting sustainable resource use and management, this paper aims to examine the efforts for collective action of ten village communities to manage its local forest resources from an institutional perspective. Drawing insights from the empirical work in 10 village communities in Orissa, that are adjacent to forest areas, the paper tries to explore the differential levels of success in the collective action efforts of these communities for ensuring sustainable resource use and management. Secondly, the paper examines the role of local level community institutions in collective action for sustainable resource management.

The paper is divided into five sections. Following the introduction, the second sections deals with larger issues of institutional and policy changes concerning natural resource management and examines the possibilities of community based approach as an alternative strategy in resource management. The third section gives a review of literature on institutions and common property management and specifies the analytical framework for the study. The fourth section empirically analyses the process of collective action for sustainable forest management and stratifies the degrees of success of the community institutions. In the conclusion, the paper discusses the role of the community institution in ensuring sustainability of the local resource base.

II. INSTITUTIONAL CHANGE FOR RESOURCE MANAGEMENT

Community based Approach as an Alternative Strategy

In the literature concerning common pool resources, ever since the publication of the Hardin's (1968) article on 'tragedy of the commons', scholars have been focusing on evolving a solution to the problems of overuse and degradation of common pool natural resources. Hardin's thesis concludes that co-users of a common pool resource are trapped in a situation in which, in the absence of any regulated access to the resource, each rational user is motivated to consume/use more and more units of the resource system till the resource is completely degraded. Thus, collective and unregulated use of common pool resource will finally lead to overuse and degradation of the resource, and individual rationality of each user will not favour any coordinated action to regulate the use of the resource. Such being the inherent problem in common pool resources, the solution, which is often put forth by Hardin and many other scholars (Demsetz, 1967; Smith, 1981; Ophuls, 1973; Hardin, 1978) is either to nationalise or privatise the resource, wherever feasible.

It is important to point out that both the solutions to the problems of common pool resource emphasise on the physical sustenance of the resource and its availability over a period of time, ignoring the needs and requirements of the users of the resource. While designing strategies for the management of natural resources, ensuring physical availability of the resource is as important as the requirements of the members of the community who make use of the resource for their daily livelihood. Further, scholars also opine that nationalisation has converts the common pool resource to a *de jure* state property, while in reality degenerating it into *de facto* open access regime (Arnold, 1998; Arnold and Stewart, 1991).

Notwithstanding the pessimism of scholarly writings on commons (e.g., Hardin, 1968), and the persistent problems of free-riding and resource degradation as a result of overuse and congestion, there has been no dearth of instances of community based collective action efforts to manage local resources successfully (see NRC, 1986; Ostrom, 1990; Berkes, 1991; Bromley et al., 1992). The case studies depicting success stories of resource management by local communities have challenged the notion that state control and privatisation are necessarily the only preferred solutions to avoid the situation of tragedy, which is so intrinsic to common pool resources. Consequently, the community based approach to management of natural resources has been acknowledged as an alternative strategy.

Why Community based Resource Management?

The shift in the policy approaches towards resource management has always been accompanied with renewed interest in the academia in studying local communities as potential resource managers, and in turn, the resurgence of community to the centre stage in development and resource management policies. Besides the state failure in achieving desirable resource management outcomes, there are a few specific reasons for the worldwide shift to community based approach as an alternative strategy for resource management.

1. Local communities, in most developing countries, share an intrinsic relation with the natural resources and continue a day-to-day interaction with the natural resource base. This is also the case with the Protected Areas and Reserved Forests, where the state enjoys absolute property rights over the resource and excludes the local people from accessing it. It is estimated

that throughout the world, perhaps half of the Protected Areas are inhabited (Borrini-Feyerabend, 1996). In India, over two-thirds of Protected Areas have human habitation, and a larger proportion have some form of human activity apart from tourism (Kothari et al., 1989). Such being the situation, exclusion of local communities from resource management is not always a desirable or possible strategy.

2. The resource management efforts which exclude the local communities often result in serious conflicts over resource use. Kothari et al.'s (1989) survey of 222 Protected Areas in India reveals that at least 47 of them had had physical clashes between people and forest officials. There are many instances where people are demanding de-reservation of the Protected Area status, since they legitimately perceive their alienation from the resources they need for survival as being unjust. Finding solutions to such conflicts points towards treating communities as equal partners in resource management.

3. Community based approach in resource management is also favoured owing to the indigenous knowledge of resource dependent communities. Local communities, especially those which have a long history of peaceful coexistence with nature, possess in-depth knowledge and experience about the resource, which, if taken into account, will be of immense value both to resource management efforts and to the community. Further, with the help of local knowledge and local people's cooperation, there is a possibility that the costs involved in natural resource management can be reduced (Kothari et al., 1989).

4. With the spread of popular democracy and decentralisation, and above all with increasing demand on people's participation, the unrepresentative policies of conservation are no longer free from criticism (see, Agrawal and Gibson, 1999). The reflections of policy changes at the larger level for deepening democracy, widening development up to the grass-roots level and empowering the indigenous population have also become visible in the natural resource sector. Empowering the local people, accelerating their development process and allowing them a greater say in the policies that affect their livelihood can be perceived only through

uplifting 'communities' and giving them a share in natural resource conservation.

5. The increasing presence of civil society and evolution of Non-governmental Organisations (NGOs) as an intermediary in development, which relies more on community than the state, also contributes to the changing status of community in the conservation debate today. The efforts of the NGOs to amplify the voices of the local indigenous population, and build their capacity for collective action at the local level for their own development have also helped the community regain its strength and be able to conserve and manage its local natural resources. Further, the flow of heavy financial aids from external donor agencies, which emphasise on local knowledge and people's participation, boost the communities to take the lead role in conservation activities with the help of local NGOs.

6. The scholarly writings about the ecological histories, environmentalism and success stories of communities managing their local resources sustainably also strengthen the position of the community as a potential resource manager. Historical ecologists emphasised the 'anthropogenic' nature of forest and the inalienable history of environments and human population (see, Denevan, 1992; Anderson and Posey, 1989). Scholars have pointed out now that there can be many occasions in which human activities and desirable levels of biodiversity can coexist (see, Arhem, 1985; Michon and Foresta, 1990; Posey, 1985; Saberwal, 1998; Western, 1989). Such possible coexistence paves the way for inclusion of local communities in strategies for effective management of these resources. Even the Indian environmental historians (see, Guha and Martinez-Alier, 1997; Guha, 2000; Gadgil and Guha, 1995) have also pointed out the peculiarity of Indian environmentalism, which lies in its inseparability from human population. Unlike northern America, where a clear distinction between natural landscape and human establishment exists, the Indian natural landscape has always coexisted with human population. Under such conditions, it becomes clear that any policy aiming at excluding local communities from conservation activities is bound to fail.

7. The advancement of research in common property has also glorified the capabilities of local communities in sustainable management of its local natural resources (see, Berkes, 1991; Ostrom, 1990; Bromley et al., 1992; McKean, 1992; NRC, 1986; Wade, 1988). These works point out towards community as a suitable alternative to nationalisation and privatisation, highlighting the importance of local knowledge and local institutional arrangements.

Recognition of the significance of community in conservation and management activities of natural resources leaves us with a further question: do all communities are equally capable for sustainable resource management? If returning back to community is the preferred solution to the problems of over-use, free-riding and degradation of natural common pool resources, then why some communities vary with respect to their capability in ensuring sustainability of the resource in their effort to manage the local resources. The answer to such questions lies in the kind of institutional arrangements that exist in the resource managing communities, since the existing institutional arrangements direct the resource management outcomes to a great extent. The following section, therefore, discusses the role of community institutions in the management of local (common pool) resources and ensuring their sustainability.

III. COMMON POOL RESOURCES, INSTITUTIONS AND SUSTAINABILITY A THEORETICAL UNDERPINNING

Defining Common Pool Resources

Before going further towards discussing the role of communities and their institutions in the management of local Common Pool Resources (CPR) through collective action, we should look at the question as to what is CPR and why CPR requires collective action at all.[1] Common pool resources can be defined as that natural or man-made resource, from which it is difficult to exclude other potential users from using the resource and such joint use by more than one user involves subtractability (Berkes and Farvar, 1991; Ostrom, 1986a). Common pool goods are distinguished from two other types of goods, i.e., pure public goods and private goods. Such classification of goods is based on changes along two analytical dimensions: exclusion of others from the resource and jointness of use. In the case of pure public goods, it becomes not only difficult but also sometimes impossible to exclude the co-users from using the resource, but such joint use does not involve subtraction of the resource. For example, while it is difficult to exclude somebody from using the services provided by a street light, at the same time one's use of such service does not necessarily affect other's use of the same service. On the other hand, private goods can be easily excluded from outsiders, but if joint use of the good begins then the resource starts to deplete. Thus, common pool resources exist in between the pure public goods and private goods, and share the characteristics of subtractability with the private good, on the one hand, and the characteristics of non-excludability with the pure public good on the other (McKean, 2000).

The twin characteristics of 'subtractability' and 'non-excludability' make the common pool resources a special kind of resource, in which it becomes difficult to protect the resource from depletion if some mechanism to exclude the non-contributing users is not devised. Proper maintenance of common pool resources requires some kind of institutional arrangements of property rights or resource management regimes through collective action to exclude the non-members or non-contributors from availing benefits from the resource. In the absence of such collective institutional efforts to address the problems of exclusion and joint use, common pool resources become *de facto* 'open access resources' accessible to anyone. To put it precisely, the management of CPR requires collective action at the community level owing to three factors. First, the good is jointly produced, or else it would not require collective action at all. Common pool resources are such that they cannot be produced in isolation by any single individual's effort. It requires joint or common efforts by the community for its production and management. Second, CPR confers benefits on all members of the group, making it impossible or impracticable to exclude members who fail to contribute for production of CPR. Third, production of benefits in CPR involves cost, which should come from all the members of the community (Heckathorn, 1993: 331).

Problems in Commons Management

Having acknowledged the necessity of the process of collective action in the production of CPR, we now move forward to discuss the challenges that resource users face in use and management of such common pool resources. In any CPR situation, the resource users mostly face two problems: problem of providing the resources (provision problem) and the problem of appropriating the resource (appropriation problem) (Ostrom, 2001). Provision problem in CPR focuses on investment for production and maintenance of the resource itself. Production of benefits from CPRs often requires long-term investments for the construction of the resource and its proper maintenance. Provision problem involves determining the type and level of regular maintenance that will sustain the resource system over time. It is important to note that in CPRs investments are usually long-term, in which costs are incurred in present while the benefits out of it are acquired in future, making it difficult for taking any decision for investment. When this difficult long-term problem is combined with free-riding of other resource users, maintenance of a resource system becomes a challenging task.

With regard to the appropriation problem in CPR environment, key focus is how to allocate a fixed amount of resource among resource users so as to avoid conflicts over the assignment of rights.[2] Equitable distribution or allocation of benefits forms an important appropriation problem in most CPR environments. In the absence of a firm rule on who can appropriate how much in a CPR environment, the situation becomes like a Prisoner's Dilemma game and free-riding becomes inherent. In such cases none will have any incentive to leave any portion of the resource for other user, which ultimately result in degradation of the resource system. However, if the access to the CPR is limited and a well-defined group of users rely on the resource jointly, the incentive of the users to appropriate will depend upon rules governing the nature of appropriation, and their enforcement and monitoring.[3]

Despite the fact that free-riding and lack of coordination poses a serious threat to collective action, several scholars have pointed out conditions required for successful collective action at the community level for the effective management of CPRs.[4] Among several other factors pointed out by these scholars, the presence or absence of 'institutions' is a prominent one, which plays a crucial role in shaping the efforts for collective action by members of the community. Ostrom's (2001) analysis also suggests that the other two problems (provision and appropriation) in the CPR context arise because of institutional problems. In the absence of institutional arrangements to design firm rules regarding access to and allocation of resource, the twin problems of provision and appropriation becomes prominent and stand in the way of collective action for sustainable management of CPRs. It is, thus, essential to clarify what 'institution' means in the first place in the context of common pool resources.

Institution as a Solution

A study of institutions, in the context of CPR, focuses on the laws and conventions of community that either directly allocate resources, or establish processes and constraints for its members to make allocative decisions. Notwithstanding the use of the term 'institution' in our day-to-day life, it is mostly used with reference to rules, regulation, and prescriptions to behave in a certain way. It is often defined as rules about behaviour, 'especially about making decisions' (Rinker, 1982: 4) and about 'individual expression and choice' (Pllott, 1979: 156). However, focusing only on the 'rules' does not serve the purpose of defining institutions properly. In her attempt to define 'institutions', Ostrom distinguishes rules from physical and behavioural laws and stresses upon the prescriptive nature of rules, which refers to 'actions that are required, prohibited or permitted; and are commonly known and used by a set of participants to order repetitive, interdependent relationships' (1986b: 5). The prescriptive nature of rules becomes visible from the fact that people who follow certain rules always know and internalise that if they violate rules, other individuals will held them accountable for that.

There has been considerable disagreement by scholars over the nature of rules. Scholars like Ganz (1971) and Shimanoff (1980) limit the scope of prescriptive rules to only 'obligation' and 'prohibition'; while on the contrary, V. Ostrom (1980) and Commons (1957) include 'permission' in their conception of rules (cited in Ostrom, 1986b). Such limitation of the scope of rules by these scholars, whether to 'prohibition or permission', arises from their very way of defining rules as prescribing a particular action to be done. Instead, Ostrom consider rules as providing a set of outcomes or actions, which it does in three ways: first, by forbidding certain action; second, by permitting certain actions or outcomes; and third, by requiring a particular action or outcome. The third type of rules expects individuals to perform a particular action and is less used in everyday life.

Ostrom defines institutions as 'a set of working rules that are used to determine who is eligible to make decisions in some arena, what actions are allowed or constrained, what aggregation of rules will be used, what procedure must be followed, what information must or must not be provided...' (Ostrom, 2001: 167). For Ostrom, working rules may not necessarily represent, be a part of, or be backed by formal laws that are designed by legislative regulations. However, this should not lead one to conclude that working rules are always formed beyond the purview of formal laws. Formal laws are important sources of working rules among several others.[5] When formal laws are known to all the participants; actively enforced and its violations are monitored, coercing those who violate them, they become working rules (Ostrom, 2001: 167-68). Thus, common knowledge, enforcement and monitoring are three important features of working rules or institutions.[6]

Arguing in a similar tone, North (1991) defines institutions as humanly devised systems that structure the interaction of its members in social, economic and political arenas. As an organised and well-defined system, it constraints some behaviours and favours or facilitates others. It enforces sanctions negatively in the form of punishments when its prescribed rules are violated and positively in the form of rewards when such rules are complied. Institutions exist both as formal entities through constitutions, laws and well-defined property rights; and as informal agents through customs, traditions, norms, codes of conduct, social taboos, etc. Institutions, thus, form the basis for human interaction and relationship among individuals within a society, consisting of both formal and informal entities. They provide a set of rules for cooperation and competition and thereby adjust conflicting claims of different members of community and of groups for scarce resources.

It has now been affirmed that the presence of institution is very much necessary to maintain common pool resources. In fact, it is the institutional arrangements, which create a property right in resources, establish management regimes; and distinguishes common property from open access resources.[7] When the designed institution restricts the access of non-members into the resource and entrusts joint property right to the users of the resource, common property regime is said to be evolved. With the presence of an institution, the property becomes inaccessible to all openly, rather the access is limited to a specific group of users who hold their right in common (Runge, 1981, 1984, 1992; Bromley and Cernea, 1989; Bromley et al., 1992).[8] Such an institutional arrangement in resource, where access to the resource is limited to specific members, which can be easily distinguished from the non-members; and where the user group or the community holds property in common and share rights and duties towards the resource, is known as common property regime.[9]

IV. COLLECTIVE ACTION, COMMUNITY INSTITUTION AND THE FOREST

The empirical work for the research was carried out in ten forest fringe villages in the Dhenkanal Sadar Block of the Dhenkanal district of Orissa. The villages were chosen purposively, based on the criteria that they lie within a distance of three kilometres from the forest, they have sufficient dependency on the forest, and the communities should have some kind of experience in collective mobilisation for management of their local resource. Case study method, supplemented by group discussions and in-depth interviews with key informants was used to elicit the necessary information regarding forest use, dependency, and efforts to its sustainability management. During the course of inquiry, information regarding social composition of the members of the village community, their past history of forest protection, and their degree of success in collective action in designing institutions for effective management of local forest resources were collected. The following table 1 indicates the names of the villages, their social composition and their experience in forest protection.

Table 1: The study villages

Name of the Village	Social composition	Since when protecting	Collective action
Gandadhar Prasad (G. Prasad)	Heterogeneous	1985	High
Jhankira	Heterogeneous	1960	High
Biradia	Heterogeneous	1993-98	None
Ambanali	Heterogeneous	1993-98	None
Kadua	Homogeneous	1993-98	None
Krushna Kumar Pur (K.K. Pur)	Heterogeneous	1993-98	None
Nagiapasi (N. Pasi)	Heterogeneous	1991	High
Korian	Heterogeneous	1996	Moderate
Krushna Prasad (K. Prasad)	Heterogeneous	1986	High
Padma Lav Pur (P.L. Pur)	Heterogeneous	1986	Moderate

Social Composition of the Villages

The social composition of the villages is very much similar with other villages of rural Orissa. All the villages except *Kadua* are multi-caste villages. The village *Kadua* is a completely tribal village situated inside the *Kapilas Reserved Forest.* One of the striking features of the remaining nine villages is that both caste and tribe coexist in these villages. People from several caste and subcastes found in Orissa, such as, Brahmin, *Karan* (record keeper), *Khandayat* (traditional warrior caste), *Chasa* (cultivator caste), *Teli* (oilman), *Tanti* (weaver), *Barika* (barber), *Keuta* (fisherman), *Gouda* (milkman) and Schedule Castes and Scheduled Tribes, reside in these villages. Social structure of the villages is very much stratified, with Brahmins at the top and Scheduled Castes at the bottom. Though the Scheduled Tribes do not come under caste system generally, but in these villages, as both caste and tribes coexist, tribals are looked down in the general social hierarchy. Agriculture is the primary occupation for most of the caste groups. The scheduled castes and scheduled tribes, who are mostly landless or possess very meagre amount of land, work as agricultural labourers in the fields of higher caste people.

Resource, Resource Dependency and Resource Protection

All the villages except *Jhankira* and *Gangadhar Prasad* (G. Prasad) are situated in the vicinity of the Kapilas Reserved Forest of Dhenkanal district. The legal status of the forest area near the village *Jhankira* and *G. Prasad* was that of village forest as per government records.[10] The forest has been an important source of livelihood for a majority of the population of the villages. Collecting fuel wood, sal leaves and other forest produces from the forest are the major occupation of the tribals in the villages. Almost all the villagers depend upon the forest for fuel wood. The poor people, mainly tribals and other lower caste people collect the fuel wood themselves, whereas the rich employ others to collect fuel wood for them on payment basis. The farmers collect small timbers from the forest to make agricultural equipments. There are certain agrarian festivals during which the farmers collect small timber to make new equipments. Collection of forest produce has been the sole source of livelihood for certain families. Particularly members from the scheduled tribes collect sal leaves, sal seeds, mahua flowers, rope making material, fuel wood and several other forest produce from which their daily expenditure comes. There are certain basket-maker families from among the schedule castes who depend upon forest for bamboo to be used as raw material in basket making. While the male members of the family collect bamboo from the forest, it is the female members who make baskets out of it. Every household depends upon forest at the time of annual repairing of the house. As most of the houses are *kachcha* houses with thatched roofs, people have to depend upon forest for bamboos, ropes and other things that are required at the time of annual maintenance.

The communities differed with each other with respect to their experience in collective mobilisation for resource protection, which ranges from 5 to 40 years. Out of the ten villages, collective efforts for forest protection did not exist in four villages, i.e. *Biradia, Ambanali, K.K. Pur and Kadua,* during the field work. In the remaining six villages, there were institutional arrangements for local resource management with different degrees of functionality. The above mentioned four villages started protecting their forests around 1993-94 and by 1998, the forest protection committees had become non-existent. Now, the forest attached to these villages is a kind of 'open access resource' and has become a major attraction for wood contractors.

The village *Jhankira* had the maximum experience of collective action efforts for forest management. The village has been protecting its 'village forest' since 1960. In the remaining five villages, where community institution existed, the protection efforts started during the mid-1980s. That was the period when community forestry was gaining momentum in Orissa. The Government of Orissa had passed its Village Forest Rules in 1985 giving the responsibility of village Forests to local communities. By 1990, the Orissa Government had passed several resolutions giving a share to local communities in the management of protected forests. Finally, in 1993, the Government of Orissa came out with its Joint Forest Management programme. Encouraged by these efforts of the government and owing to their own requirements these villages formed Forest Protection Committees to manage their patch of forests. With the formation of Protection committees, the forests, which were a sort of 'open access resource', became common property with a clear set of rules about who will enjoy the benefits from the resources. The forest protection committee started in *G. Prasad* in 1985; in *K. Prasad* and *P.L. Pur* in 1986; in *N. Pasi* in 1991 and by 1996 the village *Korian* formed such committee. The self-initiated protection committees of *K. Prasad*, *N. Pasi* and *Korian* got recognition from the local DFO's office and were registered with the Joint Forest Management (JFM) programme of the Government of Orissa. However, the village Forest Protection Committee of *G. Prasad, Jhankira* and *P.L. Pur* were functioning independently at the time of field work.

The Institutional Arrangements and the Rules in Use

The presence of community institution in the six study villages was observed to play a crucial role in ensuring sustainability of collective action efforts for local resource management. In the context of natural resource management, the institutions refer to rules about who could use what resource, when, where and how. The complete array of this rule structure is known as 'institutional arrangements', which determine the way of resource appropriation. The institutions, in the study villages, were responsible for designing rules for forest protection and for ensuring their conformity by the community members through monitoring of rule violations. Almost all the villages, where the community institutions were functioning, had two broad kinds of rules designed by their institution. They were access rules and internal use rules.

Access Rules: The institutions designed 'rules of access' to determine who will have the rights to access the resource and who will be excluded from using it. The crafting of access rules in the villages was the most significant step in their efforts for resource protection, since it is these rules which specified the property rights and created a management regime over forest. The access rules designed by the local institutions conferred exclusive property rights over the forest to the community members and excluded the other non-members (outsiders) from having such rights.

After the establishment of community institutions and formation of access rules, the communities were able to prohibit outsiders from entering inside their protected patch of forest. The Village Forest Protection Committees had demarcated the boundaries for protection. Even though it has not been possible to separate the protected patches by fencing it, yet, no doubt there existed a social fencing inside the forest area. The members of the communities know how much forestland has been entrusted to them for management and regeneration. Further, some communities have also installed stone pillars to mark off their protected patch from the rest.

Before the formation of access rules, forests were 'open access resources' in the villages. When the protection committees were not in existence in any of the villages, the forests attached to these villages were the main source of attraction for wood contractors. People from nearby localities were also using the forest for their personal gains enormously. In fact, there was a competition that time to deplete forest resources, as depletion by one user was curtailing the chance of another to do so. However, with the formation of protection committees and access rules, the villages were able to exclude others from entering into their forests. They kept a constant eye upon the forest thefts and illegal use by other communities. In a couple of years the use of the resource was regulated and only the members of the community upon which the management responsibility was entrusted became the sole users of the resource.

However, in the case of four villages, namely *Biradia, Ambanali, Kadua and K.K. Pur*, when the protection committee became extinct around 1998 the situation became grim as members from other villages started accessing the resource. It took no time to revert the earlier situation, as was in practice before the formation of the institution. The forest became again a source of income for wood contractors. Even the villagers themselves were found to cut trees heavily from the forest areas.

Internal Use Rules: The second category of rules, which the community institutions designed in the villages were related to sharing of the benefits of the resource and its proper maintenance. Once the resource was provided and property rights were determined, the next challenge for the communities was to design ways for proper maintenance of the resource and share the benefits out of it. During the course of fieldwork no internal use rules were observed in those four villages, where community institutions did not exist. In the remaining six villages, the institution had designed the following specific rules regarding internal use of the resource.

(i) *Rule Compliance:* First and foremost, the institutions designed rules for ensuring compliance from the community members and to deal with those who do not comply with. There were provisions for punishment for those who violated rules. The punishment varied from payment of fines in cash to prohibition from use of the forest produces. Usually, none violated the rules as conformity to rules brought them a share in forest produces. The benefits from the forest have been a major incentive to obey the rules of the committee. However, it is not appropriate to conclude that only economic incentives compelled the villagers to abide by the rules and contribute to the committee for forest protection. Often the social obligations and the feeling of being together prompted the villagers to pay for forest protection.

(ii) *Contribution for Collective Action:* The institutions designed specific rules regarding contributions from community members towards collective action efforts. Rules for both economic and non-economic contributions were in existence in the six villages, where the institutions were present. Members of the communities contributed both in terms of cash and kind for the purpose of management. They provided free labour whenever necessary for maintenance of the resource. Besides, in certain villages, each household contributed rupees five per month, which was invested in protection activities. The community institutions of *Nagiapasi and Gangadhar Prasad* formed rules to collect a minor charge of 50 paisa from those who collected a head load of fuel wood from the forests.

(iii) *Resource Maintenance:* The community institutions of the villages were primarily responsible for protection and maintenance of the forest patch. One of the important things to ensure resource maintenance was to check thefts and illegal entrances into the forest. The institutions had well defined rules in this regard to check forest thefts among the members of the community. In some villages, like *Nagiapasi* for example, the villagers practised *thenga pali* in the beginning to ensure protection from wood thieves and other outside appropriators.[11]

(iv) *Resource Sharing:* The community institutions took utmost care to ensure that the benefits from the resource were distributed equitably among the villagers. With prior permission from the Forest Protection Committee, the villagers were allowed to collect bamboo and minor timbers for house construction or for maintenance of rooftops of the house. The institutions designed rules to give each household an equal share from the resource pool.

(v) *Conflict Resolution:* Resolution of conflicts that arise at the time of management is the major challenge ahead of any community protecting forests.[12] The community institutions designed rules for such purposes, and tried to resolve the conflicts within the villages through the protection committee and imposed fines on the offender, which were also acceptable to the community members. However, the community faced challenges in dealing with the forest offences by the neighbouring non-members and commercial interests, as it did not have either social or juristic control over them. Such cases were usually referred to the Forest Department.

Differential Approach to Rules and Varying Degrees of Success

The presence of community institution to design rules for sustainable resource management notwithstanding, different communities showed different degrees of success in their efforts towards collective action. The differential collective action outcomes were observed to be directly related to the institutional factors. The ten study villages performed differently, since they approached the rules of the institutions in different manner. To gauge the relative success of these villages in collective action, the existence of community institutions to design rules of access and internal use was taken into consideration. The performance of villages in collective action was further divided into three categories: low, moderate and high.

(i) ***High Degree of Rule Compliance and Strong Monitoring System***: Out of the ten villages taken for the empirical survey, rules compliance and the monitoring system to punish the rule breakers were found to be strong in four villages, namely, *Gangadhar Prasad* (G. Prasad), *Jhankira, Nagiapasi* (N. Pasi), and *Krushna Prasad* (K. Prasad).

These four villages have been successful in crafting local community institutions, which could design rules regulating the access and use of the forest produces. And such rules have been backed by a strong monitoring system, which also included assigning punishments to those who fail to comply the rules. In three cases, such as in G. Prasad, Jhankira, and N. Pasi, the rules were written down in a register, below which the head of the households of the villages had put their signature. It made the point clear that rules that were created for regulating the access and use of the forest in these villages were well circulated and the members of the communities were all aware about them. Such common and shared knowledge[13] enhances the possibility of cooperation by providing 'assurance' to one that others also know the same rules and will cooperate by showing conformity to the rules.[14]

The village G. Prasad was protecting its Village Forest of 177 acres since 1985-86, and now the village institution is a JFM Committee. In the early 1980s the village forest of G. Prasad was completely deserted and the forestland was just like a barren hill track. Since that was a village forest the State Forest Department had taken little efforts towards its regeneration. Then in mid-1980s, the villagers decided to regenerate the forest and to begin with stopped allowing grazing inside the forest area. Later the villagers undertook a massive plantation programme and now the forest stands looking evergreen. Similarly, the Jhankira village was also engaged in protecting its village forest for last 40 years. This village had the longest history of forest protection from among the ten villages taken for survey, where the protection activity started in the early 1960s. Compared to these two villages, where protection activity had been carried out in Village Forest areas, the village N. Pasi and K. Prasad had taken the responsibility of protecting Reserved Forests attached to their village through JFM programme. The Reserved Forests of these two villages were highly degraded prior to the protection by the communities, and with active efforts by these two villages the protected patch is thickly grown forest now.

Protecting the resource from the free riders, who violate the rules for personal gain without contributing towards its protection, is definitely a challenging task before any community aiming to protect a forest patch. And the problem becomes more serious when the free rider belongs to another community, upon whom the resource users neither have social nor legal control. However, these four villages had been successful in prohibiting both insiders and outside members of the community from breaking rules created for forest protection. This has been possible because of the personal efforts that these community members had taken to patrol the forest areas during night times. All the four villages had developed a rotation system, and accordingly, the responsibility of guarding forest was divided among the households of the communities. Currently, N. Pasi and G. Prasad had appointed a permanent forest guard from inside the village, whose responsibility is to check the illegal entrance into the forest.

These four villages were categorised as 'High Collective Action' communities since, a local institution existed for making rules for forest protection, rules were obeyed by the members of the community and finally, a strong monitoring system existed to find out both insiders and outsiders, who tried to violate the rules.

(ii) ***Strong Rules with a Weak Monitoring System:*** Korian and Padma Lav Pur (P.L. Pur) were the two villages, among the ten, where there also existed local institutions with strong rules for forest protection. Both the villages are heterogeneous, in terms of their caste composition, with a preponderance of Tribals in the P.L. Pur village. In the other village, the social composition included households belonging to several castes and a few tribal households.

The efforts towards collective action for forest protection were a relatively new phenomenon in the Korian village, which started forest protection in the mid-1990s. In these years, the village had been successful in crafting an efficient institution, which could come out with strong rules for forest regeneration. The members of the community were also found to be very much sincere in complying the rules. However, the village institution was not able to develop monitoring systems to restrict the free riders, mostly who were members of outside community. The forest guards, that the Korian Forest Protection Committee had appointed, were able to minimize the occurrences of free riding by members belonging to their own community. However, the institution has not been successful in prohibiting members from neighbouring villages from accessing the resource and using the forest without contributing anything. In other words, the community has been facing challenges to establish a common property regime inside the forest, since members from other communities continue to access the forest, without paying anything for their use of the resource. Similar has been the case with P.L. Pur village. Although the P.L. Pur Forest Protection Committee has been able to design strong rules for forest protection and induce its members to comply with those rules, monitoring has been a problem, especially with the outside free riders. With their sincere efforts in forest protection over more than one and half decades, the forest area in P.L. Pur has regenerated substantially, which has become an attraction in the recent years for outside non-members to free ride. However, monitoring the action of these outside non-members has not been successful in P.L. Pur village.

These two villages were categorised as 'Moderate Collective Action' villages, since the communities were successful enough in establishing a complete common property regime inside their protected patch of forest, nevertheless having Forest Protection Committees and strong protection rules.

(iii) *Non-existence of Institutions*: In the remaining four communities, namely, Biradaia, Ambanali, Kadua and Krushna Kumar Pur (K.K. Pur) no functioning local institution was found during the course of fieldwork, which could design rules for forest protection. In fact, all these four communities belong to one revenue village of Krushna Kumar Pur. Even though for the government records, these four are different hamlets of one single village, yet the four communities are separate social entities having their own *jajmani* system, around which the village economy/society revolves. The hamlet Kadua is a completely Scheduled Tribes settlement, with *'Sabar'* and *'Juanga'* tribes inhabiting in it. The rest three are heterogeneous so far as their social composition is concerned.

All the four communities started the protection activities separately in the forest patches adjoining their hamlets in the year 1993. Initially, the original K.K. Pur started the protection activity and inspired by its efforts, the other three communities also established Forest Protection Committees. However, in these four communities even through the institutions created specific rules for forest protection, they were not complied by the members of the communities. Whatever rules the institutions created in the beginning, they were neither followed strictly nor the institutions were capable enough in bringing out strong monitoring systems for regulating the members' non-conforming activities concerning forest use. The village Kadua was the first one where such institution became defunct due to non-conformity of rules and non-existence of a monitoring system. Gradually, the rest three also followed the same path and within five years, the Village Forest Protection Committees in these four communities became non-existent, and along with it the efforts towards forest protection.

These four communities were categorised under 'Low Collective Action' category, since these communities failed to establish a common property regime in the forest and did not continue their efforts for forest protection.

V. DISCUSSION AND CONCLUSION

Need for Robust Institutional Arrangements

The institutional analysis of the ten studied villages reveal the fact that only presence and absence of the institution is not always sufficient for sustainable resource management, despite being a necessary condition for it. The existing institution must be a robust one with strong rules for resource appropriation and good monitoring system. Institutional arrangements for sustainable resource management at the community level must be understood as a dynamic process, which involve a continuous interaction among the community members and the designed institution. There exists a back and forth movement between these two concerning the rules for regulating the resource use. The institution formulates the rules and expects the community members to comply such rules. The rule formation should necessarily be backed by a strong and efficient monitoring system to ensure that rules are complied, and accordingly the institution can accord positive incentive in the form of rewards to those who show conformity to rules and negative incentive through punishment to those who violate them. The institutional arrangements without a strong monitoring system fail to restrict free-ride and hence, could not establish a well defined property right regime over the resource. From among the study villages, weak monitoring of rule breakings in the two communities, i.e., *Korian* and *P.L. Pur* have limited the role of the community institutions in ensuring sustainability of the forest resource (See, Table 2). The table 2 differentiates the villages based upon their institutional arrangements and depicts their differential performance in collective action.

Role of Community Institutions

The existence of institutional arrangements in six out of ten study villages was very much crucial with respect to sustainable use and management of the local forest resource. To conclude, we may deduce that the community institutions had played the following important roles, which helped certain communities succeed in collective action efforts, while others failed in it:

1. The institutions, where they existed, had ensured equitable distribution of benefits out of the common property. As discussed earlier, the institutions designed rules and regulations as to who can use the resource, when and how much resource units each member can derive. By designing rules for equitable sharing of the benefit out of the resource, the institutions solved the appropriation problem of the resource users.
2. The presence and well functioning of the institutions defined property rights over the resource and established common property regimes in the forests. When institutions were formed for forest protection, the members of the communities got property rights over the resource, which they shared among themselves within the community. Confirmation of property rights and an ensured equitable share in the benefits generated a sense of responsibility for the maintenance of the resource among the community members. With the establishment of common property regime, restriction of free-riding and proper maintenance of the resource, the community institution solved the 'appropriation problem'.

Table 2: Institutional arrangements in study villages

Name of the village	Institutional Arrangements					Performance in collective action
	Establishment of the institution	Rule formulation	Rule compliance	Monitoring		
				Insiders	Outsiders	
G. Prasad	✓	✓	✓	✓	✓	High
Jhankira	✓	✓	✓	✓	✓	High
Nagia Pasi	✓	✓	✓	✓	✓	High
K. Prasad	✓	✓	✓	✓	✓	High
Korian	✓	✓	✓	✓	×	Moderate
P.L. Pur	✓	✓	✓	✓	×	Moderate
Ambanali	✓	✓	×	×	×	Low
Biradia	✓	✓	×	×	×	Low
Kadua	✓	✓	×	×	×	Low
K.K. Pur	✓	✓	×	×	×	Low

3. With equitable sharing of benefits through the institution, all members of the community got equal access to the resource and as a matter of principle contributed for its maintenance. By giving access to the resource through property rights arrangements, guaranteeing an equal share over the benefits and demanding contributions from the users of the resource; the institution ensured the participation of all members of the community in the management of the resource. When the members visualised the benefits of collective efforts they participated actively by physically guarding the forests from any kind of offence.
4. Finally, with greater participation, guaranteed share in the resource, demand for contribution for maintenance, the institution worked towards sustainability and effective management of the newly evolved common property regimes in the villages.

The diagram shown below depicts the above mentioned roles of the community institution in collective action for sustainable common pool resource management.

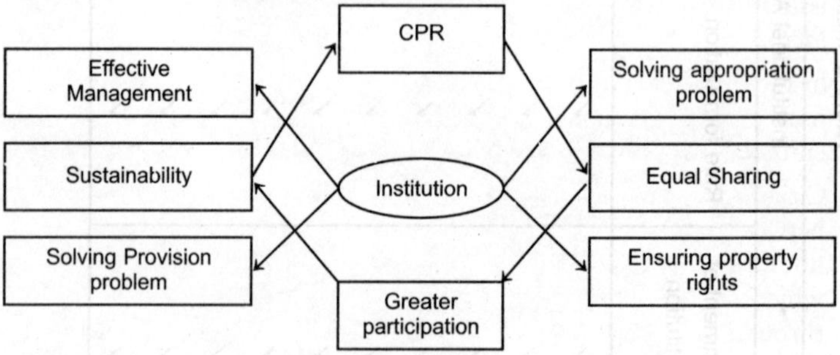

NOTES

1. The term 'collective action' in social science literature has been widely used by different scholars to understand different situations. However, the element, which is common to all, is 'mutual interest and the possibility of benefit from coordinated action'. Situations requiring collective action take several forms, one of which is the management of common property resources like fisheries, forests

and irrigation. Although sociological literature on the subject is relatively small and new (e.g. Heckathorn, 1988, 1989, 1990, 1991, 1992; Oliver, 1980; Marwell and Oliver, 1984, 1993; Oliver, Marwell and Teixeira, 1985), research on the subject by economists (e.g. Samuelson, 1954; Bator, 1958; Head, 1974; Olson, 1965), political scientists (e.g. Chamberlin, 1974; Frohlich and Oppenheimer, 1970; Hardin, 1971) and psychologists (Kelley and Grzelak, 1972; Bonacich et al., 1976; Dawes, McTarish and Shaklee, 1977) is much too large and old.

2. Resource appropriation in CPR means withdrawal of resource units by resource users from the whole resource system. Resource users may appropriate resource units for self-consumption, for using it as an input for production process, e.g. water for agricultural development; or for the market (Ostrom, 2001: 142).
3. In the case of forest management, while the provision problem relates to regeneration and maintenance of the forest area; the appropriation problems are related with equitable distribution of forest produces among the forest users.
4. For a detailed understanding of the factors essential for collective action see Wade, 1988; Ostrom, 1990; Baland and Platteau, 1996.
5. On certain occasions there arise differences between formal laws and working rules. In such cases, working rules may provide *de facto* rights over a resource that are contrary to *de jure* rights provided by formal laws. The whole discourse of legal pluralism addresses these issues (see, Rout, 2005). However, Ostrom concentrates only on *de facto* rights that are actually used in CPR settings.
6. In fact, prior to Ostrom, Commons (1957) used the term 'rules in use' or 'working rules', which are those actually used, monitored and enforced when individuals make choices of taking certain actions.
7. Ciriacy-Wantrup and Bishop (1975) makes a further distinction between an open-access CPR, in which any firm property right does not exist and none claims such rights, and a closed-access CPR, in which a well defined group enjoys property rights over the resource through their institution.
8. McKean considers that when a common property is combined with a well-defined institution, user group and property rights, it becomes a 'shared private property' (see, McKean, 2000).
9. Common property regime is distinguished from other three kinds of resource management regimes: state property regime, private property regime and open access. In the case of state property regime, the property rights (ownership) over the resource remains with the state, which controls it by its agencies like government. The resources

under private property regimes are owned, managed and used by private individuals. Open access is a situation where property rights arrangement does not exist at all and hence, none (neither individual/s nor community) can claim exclusive right over the resource.

10. Property rights over the forestland determine the difference between different types of forests and accordingly, their legal status. In Orissa, forests are divided into three categories: Reserved Forest, Protected Forest and Village Forest. While in the case of Reserved Forest, the Forest Department enjoys property rights over the forestland, in Protected Forests, the management rights and responsibilities of the forest are conferred on Forest Department, whereas the property rights on land remain with Revenue Department. The management responsibility of the Village Forests is supposed to be vested with the local communities as per the Government of Orissa Village Forest Resolution, 1985.
11. *Thenga pali* (which can be translated as stick rotation in English), a voluntary monitoring system, in which a wooden stick, *thenga*, their symbol of protection, is passed on from house to house making a symbolic transfer of responsibility of forest protection. Every household after fulfilling the task of watching, passes on the stick to the next household, whose turn it is to watch the forest the following night. The household that receives the stick is responsible for getting members of the other four households to join them when they proceed to the forest.
12. There are mainly four potential sources of conflict that may arise in any community managing forests: (i) conflict among the community institution's members, (ii) conflict with neighbouring non-members, (iii) conflict with other external commercial and industrial agents, and finally, conflict with the state, primarily with the forest department (Sarin, 1996).
13. Elinor Ostrom (1990, 2001) also regards common knowledge as one of the important characteristics of 'working rules'.
14. The Game Theoretic Approach deals with the 'problem of assurance' in a common pool situation, and discusses how the resource users become assured of each others' cooperation. For details, see, Axelrod (1981), Burdhan (1993) and Rasmussen and Meinzen-Dick (1995).

REFERENCES

Agrawal, A. and C.C. Gibson (1999). "Enchantment and Disenchantment: The Role of Community in Natural Resource Conservation." *World Development*, Vol. 27 (4), 629-49.

Anderson, A. and D. Posey (1989). "Management of the Tropical Scrub Savanna by the Gorotire Kayapo of Brazil." *Advances in Economic Botany,* Vol. 7, 159-73.

Arhem, K. (1985). *Pastoral Man in the Garden of Eden: The Massai of the Ngorongoro Conservation Area, Tanzania,* University of Uppsala, Uppsala.

Arnold, J.E.M. (1998). *Managing Forests as Common Property,* FAO Forestry Paper No. 136, Food and Agricultural Organisation, Rome.

Arnold, J.E.M. and W.C. Stewart (1991). *Common Property Resource Management in India,* Tropical Forestry Paper No. 24, Oxford Forestry Institute, Oxford, UK.

Axelrod, R. (1981). "The Emergence of Cooperation among the Egoists." *American Journal of Political Science Review,* Vol. 75 (7), 306-18.

Baland, Jean-Marie and Jean-Philippe Platteau (1996). "Halting Degradation of Natural Resources: Is There a Role for Rural Communities?" Oxford University Press, Oxford.

Bardhan, P. (1993). "Analytics of the Institutions of Informal Cooperation in Rural Development." *World Development,* Vol. 21 (4), 633-39.

Bator, F.M. (1958). "The Anatomy of Market Failure." *Quarterly Journal of Economics,* Vol. 72, 351-79.

Berkes, F. (ed.) (1991). *Common Property Resources: Ecology and Community-based Sustainable Development,* International Books Distributors, Dehra Dun.

Berkes, F. and M.T. Frvar (1991). "Introduction and Overview." F. Berkes. (Ed.). Common Property Resources: *Ecology and Community-based Sustainable Development,* International Books Distributors, Dehra Dun.

Bonacich, P., G.H. Shure, J.P. Kahan and R.J. Meeker (1976). "Cooperation and Group Size in N-Person Prisoners' Dilemma." *Journal of Conflict Resolution,* Vol. 20, 687-706.

Borrini-Feyerabend, G. (1996). *Collaborative Management of Protected Areas: Tailoring the Approach to the Context,* IUCN, Gland.

Bromley, D.W. and Michael M. Cernea (1989). "The Management of Common Property Natural Resources: Some Conceptual and Operational Fallacies." World Bank Discussion Papers No. 57, Washington DC.

Bromley, D.W.; David Feeny; Margaret A. McKean; Pauline Peters; Jere Gilles; Ronald Oakerson; C. Ford Runge and James Thomson (ed.) (1992). *Making the Commons Work: Theory, Practice and Policy,* ICS Press, San Francisco.

Chamberlin, J. (1974). "Provisions of Collective Goods and a Function of Group Size." *American Political Science Review,* Vol. 68, 707-16.

Ciriacy-Wantrup, S.V. and R.C. Bishop (1975). "Common Property as a Concept in Natural Resource Policy." *Natural Resource Journal,* Vol. 15, 713-27.

Commons, J.R. (1957). *Legal Foundations of Capitalism,* University of Wisconsin Press, Madison.

Dawes, R.M., J. McTarish and H. Shaklee (1977). "Behaviour, Communication and Assumptions about Other People's Behaviour in a Common Dilemma Situation." *Journal of Personality and Social Psychology,* Vol. 35, 1-11.

Demsetz, H. (1967). "Towards a Theory of Property Rights." *American Economic Review,* Vol. 57 (May), 347-59.

Denevan, W.M. (1992). "The Pristine Myth: The Landscape of the Americas in the 1942." *Annals of the Association of the American Geographers,* Vol. 82 (3): 369-85.

Frohlich, N. and J. Oppenheimer (1970). "I Get it with a Little Help from My Friends." *World Politics,* Vol. 23, 104-20.

Gadgil, M. and R. Guha (1995). *Ecology and Equity: Use and Abuse of Nature,* Penguin Books, Middlesex.

Ganz, J.S. (1971). *Rules: A Systematic Study*, Mouton Press, The Hague.

Guha, R. (2000): *Environmentalism: A Global History,* Oxford University Press, Delhi.

Guha, R. and J. Martinez-Alier (1997): *Varieties of Environmentalism: Essays in North and South,* Oxford University Press, Delhi.

Hardin, G. (1968). "The Tragedy of the Commons." *Science*, Vol. 162, 1243-48.

Hardin, G. (1978). "Political Requirements of Preserving our Common Heritage." H.P. Bokaw (ed.), *Wildlife and America,* Council of Environmental Quarterly, Washington D.C.

Hardin, R. (1971). "Collective Action as an Agreeable *N*-Persons' Dilemma." *Behavioural Sciences,* Vol. 16, 472-81.

Head, J.G. (1974). "Public Goods and Public Expenditure." Duke University Press, Durham, N.C.

Heckathorn, D.D. (1988). "Collective Sanctions and Emergence of Prisoner's Dilemma Norms." *American Journal of Sociology,* Vol. 94, 535-62.

Heckathorn, D.D. (1989). "Collective Action and the Second Order Free Riders' Problem." *Rationality and Society,* Vol. 1, 78-100.

Heckathorn, D.D. (1990). "Collective Sanctions and Compliance Norms: A Formal Theory of Group Mediated Social Control." *American Sociological Review,* Vol. 55, 366-84.

Heckathorn, D.D. (1991). "Extension of Prisoner's Dilemma Paradigm: The Altruist's Dilemma and Group Solidarity." *Sociological Theory*, Vol. 9, 34-52.

Heckathorn, D.D. (1992). "Collective Sanctions and Group Heterogeneity: Cohesion and Polarisation in Normative Systems." Ed Lawler (ed.) *Advances in Group Theory and Research*, Vol. 9, JAI Press, Greenwich Conn.

Heckathorn, D.D. (1993). "Collective Action and Group Heterogeneity: Voluntary Provisions versus Selective Incentives." *American Sociological Review*, Vol. 58, 329-50.

Kelley, H.H. and J. Grzelak (1972). "Conflict between Individual and Common Interests in an *N*-Person Relationship." *Journal of Personality and Social Psychology*, Vol. 21, 190-97.

Kothari, A., R.V. Anuradha and N. Pathak (1998). "Community Based Conservation: Issues and Prospects." A. Kothari, N. Pathak, A.V. Anuradha and B. Taneja (ed.), *Communities and Conservation: Natural Resource Management in South and Central Asia*, Sage Publications, New Delhi.

Kothari, A., P. Pande, S. Singh and D. Variava (1989). *Management of National Parks and Protected Areas: A Status Report*, Indian Institute of Public Administration, New Delhi.

Marwell, G. and P. Oliver (1984). "Collective Action Theory and Social Movements Research." *Research in Social Movements, Conflicts and Change*, Vol. 7, 1-27.

Marwell, G. and P. Oliver (1993). "The Critical Mass in Collective Action: A Micro-Social Theory." Cambridge University Press, Cambridge.

McKean, M.A. (1992). "Success on the Common: A Comparative Examination of Institutions for Common Property Resource Management." *Journal of Theoretical Politics*, Vol. 4 (3): 247-82.

McKean, M.A. (2000). "Common Property: What is It, What is It Good for, and What Makes It Work." Clark C. Gibson, M.A. McKean and E. Ostrom (ed.) *People and Forests: Communities, Institutions and Governance*, The MIT Press, Cambridge.

Michon, G. and H. de Foresta (1990): *Complex Agro-forestry Systems and the Conservation of Biological Diversity*, Proceedings of the International Conference on Tropical Biodiversity in Harmony with Nature, Kuala Lumpur (Malaysia), 12-16 June 1990.

North, D.C. (1991). "Institutions." *Journal of Economic Perspective*, Vol. 5 (1): 97-112.

NRC (1986). *Proceedings of the Conferences on Common Property Resource Management*, National Academy Press, Washington DC.

Oliver, P. (1980). "Rewards and Punishment as Selective Incentives for Collective Action: Theoretical Investigation." *American Journal of Sociology,* Vol. 85 (6), 1356-75.

Oliver, P., G. Marwell and R. Texereira (1985). "A Theory of Critical Mass. I: Interdependence, Group Heterogeneity and the Production of Collective Action." *American Journal of Sociology,* Vol. 91(3): 522-56.

Olson, M. (1965). *The Logic of Collective Action: Public Goods and the Theory of Groups*, Harvard University Press, Cambridge.

Ophuls, W. (1973). "Leviathan or Oblivion." H.E. Daly (ed.), *Towards a Steady State Economy,* W.H. Freeman, San Francisco.

Ostrom, E. (1986a). "Issues of Definition and Theory: Some Conclusions and Hypothesis." *Proceedings of the Conference of Common Property Resource Management,* National Academy Press, Washington, DC, 597-615.

Ostrom, E. (1986b). "An Agenda for Study of Institutions." *Public Choice,* Vol. 48, 3-25.

Ostrom, E. (1990). *Governing the Commons: The Evolution of Institutions for Collective Action*, Cambridge University Press, New York.

Ostrom, E. (2001). "An Institutional Approach to the Study of Self-organisation and Self-governance in CPR Situations." Ulaganathan Sankar (ed.), *Environmental Economics,* Oxford University Press, New Delhi.

Ostrom, V. (1980). "Artisanship and Artifact." *Public Administration Review,* Vol. 40 (4), 309-17.

Pllott, C.R. (1979). "The Application of Laboratory Experimental Methods to Public Choice." C.S. Russell (ed.), *Collective Decision-Making: Application from Public Choice Theory,* John Hopkins University Press, Baltimore, MD, 137-60.

Posey, D. (1985). "Indigenous Management of Tropical Forest Ecosystems: The Case of Kayapo Indians in the Brazilian Amazon." *Agroforestry Systems,* Vol. 3, 139-58.

Rasmussen, L.N. and Ruth Meinzen-Dick (1995). "Local Organisations for Natural Resource Management: Lessons from Theoretical and Empirical Literature." EPTD Discussion Paper No. 11, International Food Policy Research Institute, Washington, DC.

Rinker, W.H. (1982). "Implications from the Disequilibrium of Majority Rule for the Study of Institutions." P.C. Ordeshook and K.A. Shepsle (ed.), *Political Disequilibrium,* Kluwer-Nijhoff Press, Boston. Originally published in the *American Political Science Review,* Vol. 74, 1980, 432-47.

Rout, S. (2005). "Whose Forest is It? Customary Rights and Legal Pluralism in Joint Forest Management in Orissa." *Indian Socio-Legal Journal,* Vol. 31 (Special Issue on Legal Pluralism in India).

Runge, C.F. (1981). "Common Property Externalities: Isolation, Assurance and Resource Depletion in a Traditional Grazing Context." *American Journal of Agricultural Economics,* Vol. 63, 595-606.

Runge, C.F. (1984). "Strategic Interdependence in Models of Property Rights." *American Journal of Agricultural Economics,* Vol. 66, 807-13.

Runge, C.F. (1992). "Common Property and Collective Action in Economic Development." Daniel W. Bromley (ed.), *Making the Commons Work: Theory, Practice and Policy,* Institute for Contemporary Studies Press, San Francisco, 17-39.

Saberwal, V. (1998). "Pastoral Politics, Gaddi Grazing, Degradation, and Biodiversity Conservation in Himachal Pradesh, India." *Conservation Biology,* Vol. 10 (3), 741-49.

Samuelson, P.A. (1954). 'The Pure Theory of Public Expenditure', *Review of Economics and Statistics,* Vol. 36, 387-89.

Sarin, M. (1996). "From Conflict to Collaboration: Institutional Issues in Community Management." Mark Poffenberger and Betsy Mc Gean (ed.), *Village Voices, Forest Choices: JFM in India,* Oxford University Press, Delhi, 165-209.

Shimanoff, S.B. (1980). *Communication Rules: Theory and Research,* Sage Publications, Beverly Hills.

Smith, R.J. (1981). "Resolving the Tragedy of the Commons by Creating Private Property Rights in Wild Life." *CATO Journal,* Vol. 1, 439-68.

Wade, R. (1988). "Village Republics: Economic Conditions for Collective Action in South India." Cambridge University Press, Cambridge.

Western, D. (1989). "Conservation without Parks: Wildlife in Rural Landscapes." D. Western and M.C. Pearl (ed.), *Conservation for the Twenty-first Century,* Oxford University Press, New York.

Impact of Coal-fired Thermal Power Plants on Agriculture: A Case Study of Chicku (Sapota) Orchards at Dahanu, Maharashtra

9

P.R. Arun, P.A. Azeez, and *V. Mahajan Maya*

1. Introduction

Salubrious forests and luxurious orchards dominate the landscape of Dahanu taluka, in the coastal region of northern Maharashtra. It has the distinction of being one of the very few last surviving coastal green havens amidst a highly denuded and industrialised neighbourhood. In spite of being a declared eco-fragile area under Dahanu notification of 1991 (Ministry of Environment and Forests, Govt. of India) the Dahanu taluka continue to face various environmental challenges threatening the very sustenance of the natural ecosystem apart from severely affecting the livelihoods of the local people and their economy. Especially the fruit and fishery production from the area has been reportedly experiencing a mysterious and precipitous decline during the past few years. The sapota plantations along the coastal belt of Dahanu taluka, which were healthy and yielding prolifically until a few years back are yielding significantly less during the recent years. A slight decline in yield perceived by the mid-1990s had become widespread in the Taluka by the end of that decade. The decline has intensified further during the recent past affecting the local economy completely.

The precipitous decline in yields from orchards and fishery sectors has affected not only farmers and fishermen, but also people from other walks of life (such as farm labourers, transporters, businessmen and Auto/taxi workers). Vast majority of the people here are directly or indirectly dependent on the orchards and fisheries sectors for their livelihoods and are severely affected from the plummeting yields and productivity in these sectors. Dahanu's people and economy are struggling to cope with the difficult times. Immediate and serious mitigatory measures are imperative to save, this environmentally and ecologically precious fragile costal zone of Maharashtra before it succumb to the environmental challenges.

2. The Present Study

The present study trace the trend of decline along multiple dimensions and identify the probable causes and also examines general ecological concerns of the area. It examines the problems of the Sapota (Chicku) orchards from an ecological perspective, on the basis of ground level information collected from the orchards of the area. The present short-term study was taken up with two major objectives:

(1) Assess the nature and extent of the yield decline in the Sapota orchards of the Dahanu area and

(2) Make a preliminary evaluation of the role of various potential factors that are possibly affecting the Sapota yield including the emissions from the thermal power plant.

The methodology followed for the study included an extensive questionnaire survey carried out among the farmers of Dahanu during January to March 2004 using a custom made questionnaire (Appendix 1). Primary information was also collected through discussions with various people concerned, such as farmers, wholesale dealers of fruits, social workers, journalists and farm labourers. Data on factors, such as age of the plants, farming practices, flowering and fruiting patterns, water availability; irrigation methods, pest incidence and annual yields were recorded from the orchards. As it has been reported from elsewhere, other possible reasons such as a new pest (Horvitz and Schemske 2002) or a possible decline of natural enemies/predator populations of the pest species (Parrotta et al. 2002) were also examined during the study. Stratified random sampling method was the sampling strategy followed during the study. 116 farms at various locations in Dahanu taluka were selected randomly and data was collected on various ecological aspects in a structured manner following a questionnaire designed for the study (Appendix 1). Apart from the primary data thus collected, various secondary data sources were also consulted wherever necessary.

Geographic co-ordinates (Latitude and longitude) of the farms were recorded using a hand held GPS (Global Positioning System) except for a few farms that were not visited personally. Since the Dahanu Thermal Power Plant (DTPP) is the most important source of environmental pollution at Dahanu, the radial distance of various farm locations from the chimney of DTPP was also recorded (Appendix 2). The mean geographic location of orchards surveyed from various places and their aerial distances from DTPP are given in Table 1.

Table 1: Major locations in Dahanu taluka surveyed during the study

Place	Latitude* (degrees)	Longitude* (degrees)	Distance from TPP (in km) & Direction
Agwan	19.94	72.79	4.81 E, SE
Ashagad	19.94	72.88	14.10 E
Aswali	20.09	72.77	15.42 N
Bordi	20.11	72.77	16.79 N
Borigaon	20.11	72.76	18.10 N
Chikla	20.07	72.71	12.90 N
Ganjad	19.95	72.84	9.94 E
Gholvad	20.09	72.74	14.68 N
Jamshet	19.99	72.82	9.21 NE
Jhai	20.09	72.73	18.70 N
Kainad	20.02	72.78	8.10 NE
Kankradi	20.00	72.75	4.90 N
Khanoda	20.08	72.78	14.05 N
Kosbad	20.05	72.74	11.00 N
Masoli	20.00	72.74	6.08 N
Pale	19.90	72.78	7.00 SE
Saravali	19.99	72.77	4.32 N, NE
Vaki	20.05	72.75	10.50 N
Vangaon	19.87	72.73	9.05 S, SW

* Based on the mean location of the surveyed farms.

3. Ecological Significance of Dahanu Taluka

Dahanu is one of the 13 talukas coming under the district of Thane in Maharashtra. It consists of 174 villages and the Dahanu Municipal Corporation (DMC) formed in 1985 merging seven villages. Located in the northern most part of the West Coast of Maharashtra, Dahanu extends over 960 km^2 with forests covering more than half of its land area (Dewan and Chawla 1999). The eastern half of Dahanu, otherwise called 'junglepatti', is a tribal dominated hilly area with rich forest cover. The western half is

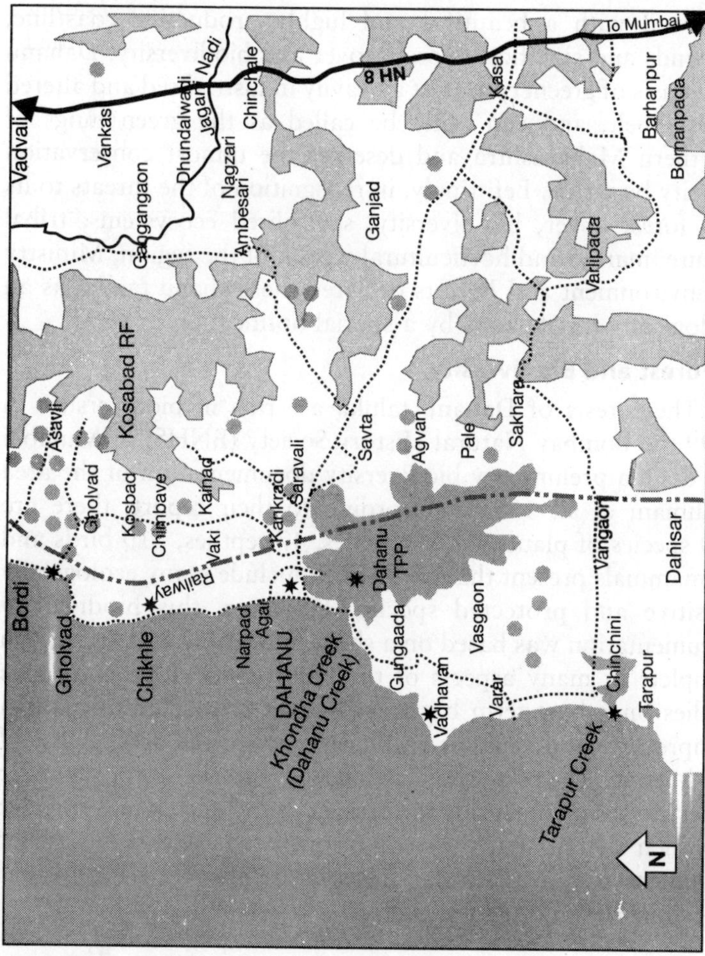

Figure 1: Map of the area showing the location of studied orchards (as faded spots)

known as 'bandarpatti' and is mainly comprised low lying coastal zone, dominated by orchards and other agricultural lands.

Dahanu is a crucially important ecological system which still sustain the relict of once continuous Western-Ghat forests existed all along Maharashtra's western boundary. Dahanu has the distinction of majority population being tribals who still depend heavily on the local ecosystem resources and traditional occupations such as agriculture and fishing for their day-to-day needs. Dahanu also an important source of fruits, vegetables, fish, fodder, milk and flowers for the near by urban areas. It is endowed with a beautiful and highly productive coastline, wetlands and significant forest cover and biodiversity. Dahanu is an oasis of greenery amidst a heavily industrialised and altered coastal belt, and can aptly be called as the green lungs of Northern Maharashtra and deserves the utmost conservation priority by virtue. Befittingly, in recognition of the threats to its rich forest cover, bio-diversity, specialised ecosystems, tribal culture, marine and horticultural wealth of the region, Ministry of Environment and Forests declared the Dahanu taluka as an ecologically fragile zone by a special notification in 1991.

4. Forest and Biodiversity

The forests of Dahanu taluka are rich in biodiversity. In 1999 the Bombay Natural History Society (BNHS) of Mumbai has done a preliminary biodiversity documentation of the area (Rahmani et al. 1999). According to their report, there are 420 species of plants, 32 butterflies, 47 reptiles, 210 birds and 30 mammals present this area, which include many ecologically sensitive and protected species. However, this biodiversity documentation was based on a short-term study and is far from complete in many aspects of the biodiversity. Further serious studies on a long-term basis need to be conducted to make a comprehensive document on biodiversity of the area. A list of 50 butterflies from the area based on the opportunistic observations made during the present study is given in Appendix 7, which signals that the area is much richer than the earlier documentation suggests, and emphasise the need for further investigations and documentation.

5. Climate

Climate of Dahanu is humid tropical with prominent southwest monsoon showers during the period of June-August. The area receives an annual rainfall of around 1800 mm (Figure 2). Although there was slight reduction in the rainfall during the recent past, there was no any drastic change in the rainfall pattern as such (See Appendix 3 for Monthly rainfall pattern over the years). Predominant wind direction during the non-monsoon seasons was from north-east and during monsoon months it was generally from south-west. According to the local farmers, the temperature regime has changed over the recent years with rise in the winter and night temperatures.

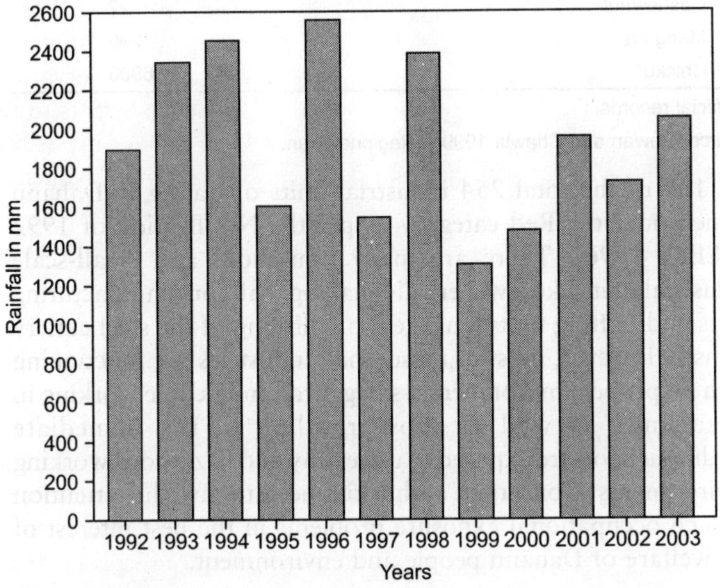

Figure 2: Annual rainfall of Dahanu during the twelve years

6. People and Economy

A vast majority of around 3 lakh strong population of Dahanu taluka live in its rural outskirts. Scheduled tribes dominate population (67%) of the taluka. Agriculture and fishery sectors together form the backbone of the economy and livelihoods of the people. A comprehensive account of the socio-economic profile of the region is given in the book by Ritu Dewan and Michelle Chawla (1999). The land use pattern of the Taluka as per the official records is given in Table 2.

Table 2: Area under different crops (Dahanu Taluka)

Crop	Area in hectars
Rice	16386
Wheat	99
Moong	250
Val	149
Chillies	150
Other Spices	170
Coconut	80
Cashewnut	22
Mangoes	290
Chicku*	6900

* Official records.

Source: Dewan and Chawla 1999 & Regional Plan.

109 of the total 254 industrial units operating in Dahanu come under the Red category as per the Notification of 1992 (NEERI 1996). There are many household and small-scale industrial units like jewellery die-making, balloon manufacturing units and buffing units (engaged in polishing of the steel cutlery items). However, most of these small industries are functioning with no proper environmental safeguards and people working in these units as well as those residing in the immediate neighbourhood are exposed to unhealthy and hazardous working environments. Concerned authorities need to give due attention to such occupational exposure problems in the best interest of the welfare of Dahanu people and environment.

7. Orchards

Orchards are the major employment source for the predominantly tribal workforce of Dahanu. Although Sapota is the dominant orchard tree in the farms of Dahanu, several other tree species such as Coconut, Mango, and Litchi are also grown in these orchards. Associated activities such as apiculture, medicinal plant cultivation and vermicomposting are also pursued in many farms. Most of the orchard farmers still follow the traditional and environmentally benign organic farming practices. Around 75 per cent of the farms surveyed during the present study were found following biological farming practices without using chemical pesticides and fertilizers. However, of late, forced by the declining yields and the propaganda by agrochemical industry, some of the farmers have deviated from their eco-friendly traditional methods and have gone for expensive and environmentally hazardous chemical practices.

Currently most orchard farms of Dahanu are reeling under multi-faceted problems. Pest menace in the farms has greatly escalated. The African Snails (*Achatina* [*Lissachatina*]) *fulica* Bowdich) which once devoured the greenery of the orchards of Bordi area few years back has been efficiently controlled by the farmers using just the common salt. Eriophyid mites have taken a heavy toll on the coconuts, which earlier had wreck havoc in the coconut farms of southern states, especially Kerala. However, various control agents have been experimented against this pest including a biological control agent *Hirsutella thompsonii* (Gopal and Gupta 2001). However, there are efficient pest management strategies have succeeded in managing this scourge to a great extent. The ICAR centre at Kosabad is conducting the final field tests of a mite management strategy tailor-made for the local environmental conditions of Dahanu. Presently the borer pests (*Trymalitis margarius and Anarsia achrasella*) are a severe problem in the Chicku orchards.

There was yield decline in Mangoes too in recent years. However unlike Sapota, which yields almost all through the year, the Mangoes are known to yield irregularly and hence, the farmers are not that unhappy with the yield decline from Mango orchards. Since the mango orchards were abundant in flower during early this year when the field surveys for the present study was under way, and the farmers were looking forward for a bumper yield next season. However their dreams were shattered overnight when the flowers charred out before their eyes due to smoggy weather in last March. From the preliminary observations it appears that the condensation of acidic dew on the flowers has caused this damage.

8. Sapota (*Manilkara achras* [Mill.] Forberg)

The Sapota (*Manilkara achras* [Mill.] Forberg), known as 'Chicku' in local language, is a major tropical orchard plant widely cultivated in Dahanu and its neighbourhoods. This tropical American fruit tree is believed to be introduced here by the Portuguese way back in 15th century (Dewan and Chawla 1999). Sapota, cultivated for its edible fruits in India is cultivated in many other tropical countries like Mexico, Guatemala and Venezuela mainly for extraction of the edible gum from the bark (Chicle gum). In India the plant is not known to be used for chicle extraction and is cultivated only for the fruits. Although sapota is only a minor fruit among other Indian fruits, India is the largest producer of Sapota in the world (Bal 1996). The seeds and bark of the Sapota tree have medicinal properties.

Sapota fruits are rich in nutrients (Table 3). It contains polyphenols and tannin along with sugars in varying concentrations depending mainly on the stage of ripening (deBrito and Narain 2002). Drying is the major processing method used for preservation for Sapota (Ganjyal et al. 2003), and many of the enterprising farmers of Dahanu have devised ingenious methods and equipments to hygienically desiccate and preserve the fruits. However attempts to exploit the marketing potentials of the processed fruits appear to be lacking.

Table 3: Nutritional Composition of ripe Sapota fruit

Constituent	Amount (per 100 gm of edible portion)
Moisture	73.7 g
Carbohydrate	21.4 g
Protein	00.7 g
Fat	01.1 g
Calcium	28.0 mg
Phosphorus	27.0 mg
Iron	02.0 mg
Ascorbic acid	06.0 mg

Source: Sulladmath et al. 1990.

The first commercial Sapota cultivation from Maharashtra was taken up in Gholvad area in 1898 (Sulladmath and Reddy

1990; Bari 2002). Since then the Chicku plantations were flourishing well in the warm and humid climate of Dahanu. Unlike many other plant species the Sapota plant has the unique characteristic of flowering and fruiting all through the year, which helped making this plant a hot favourite of many a Horticulturists. Several traditional varieties of Sapota are available for cultivation. Major cultivars include Kalipatti, Cricket ball, Chhatri, Culcutta round, and Pala varieties. Apart from these, superior hybrid varieties are also available which are developed through plant breeding techniques in different parts of the country (Eg: CO-1, CO-2 and PKM-1 from Tamil Nadu Agricultural University, Coimbatore). However these hybrid varieties are not popular among the farmers of Dahanu, and the traditional Kalipatti remains the most widely cultivated variety in Dahanu. The Plants grown in the orchards of Dahanu are Kalipatti variety grafted on the rootstock of Rayan or Khirnee (*Manilkara hexandra* or *Mimusops hexandra*). Mahua (*Madhuca latifolia*) is another potential rootstock plant for Sapota. However, *Manilkara hexandra* is the most vigorous and productive rootstock plant widely used in the orchards here.

Traditionally Sapota was considered to be a hardy species that need little attention from the farmers, as there were no serious problems of pests or diseases in the orchards. However, presently the scenario has changed and there are increased levels of pest attacks and diseases in the farms. In the young orchards intercrops such as vegetables are grown along with the young Sapota plants, while, older plantations hardly allow any undergrowth due to the thick canopy cover. Although Sapota trees yield all through the year, at Dahanu, it has two major harvesting seasons, winter and summer.

9. The Yield Decline

Of late there had been reports of extensive decline in the yield of Sapota from the orchards of Dahanu (Bari and Bari 2004). According the data obtained during the present study, the annual yields from the farms were found steadily declining over the years (Figure 3). The orchards have experienced around 50 per cent reduction in their annual yields during the past five

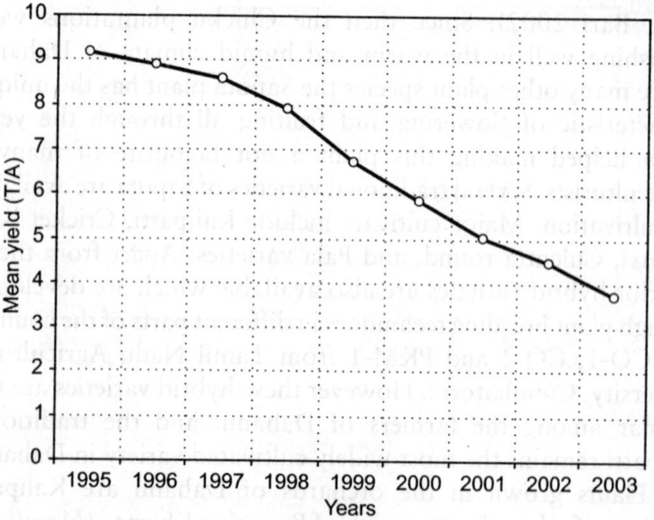

Figure 3: Mean yield of Sapota (tonnes/acre) during the past nine years

years. Although, exact time and cause of the decline is unknown, as per the data collected during the present study, earliest record of the decline was in 1995. However, the vast majority of the farms under the present study felt the brunt much later, by around 1999-2000 [Figure 4; Bars indicate cumulative number of orchards that experienced the decline (n = 116)]. Unfortunately, there had been no attempt in the past to quantify monitor or document this decline. However, few of the farmers in Dahanu have meticulously kept regular long-term record of the yield from their farms. In the present study few of such secondary data were consulted for the purpose of understanding the magnitude and temporal pattern of the decline.

According to the Wholesale traders, the Sapota pulling from the Dahanu area during the peak seasons has dropped sharply over the past decade (from around 70 truck loads in 1995 to around 20 truck loads in 2003). The magnitude of decline is far more severe in Dahanu (Figure 3) compared to other nearby Chicku growing areas such as Amalsad (located north of Dahanu in Gujarat state), where there is hardly any serious decline during the same period (Figure 5; based on the data kindly

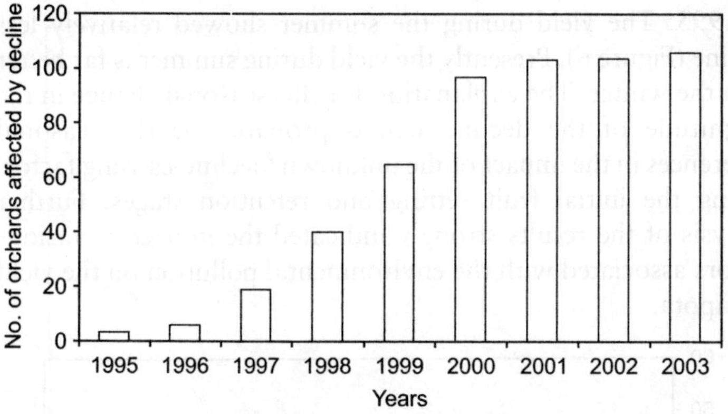

Figure 4: Temporal pattern in number of farms affected with the yield decline (n = 116)

provided by Hemantbhai B. Naik, Secretary V.V.K.S.K.M. Ltd., Amalsad, Appendix 5).

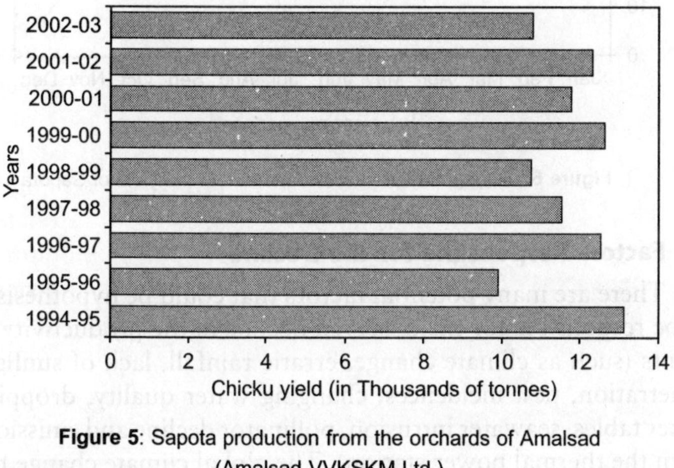

Figure 5: Sapota production from the orchards of Amalsad (Amalsad VVKSKM Ltd.)

9.1. *Seasonal pattern*

Generally the major Chicku harvesting season of the area were the winter (post monsoon) and summer. As it could be observed from the detailed month-wise data obtained from one of the orchards at Saravali area the yield during the winter season has reduced significantly during 2003 compared to that of 1995. The yield during the summer showed relatively less decline (Figure 6). Presently the yield during summer is far better than the winter. The explanation for this seasonal change in the magnitude of the decline would probably be the seasonal differences in the impact of the unknown 'decline-causing factor' during the initial fruit setting and retention stages. Further analysis of the results strongly indicated the impact of various factors associated with the environmental pollution on the yield of Sapota.

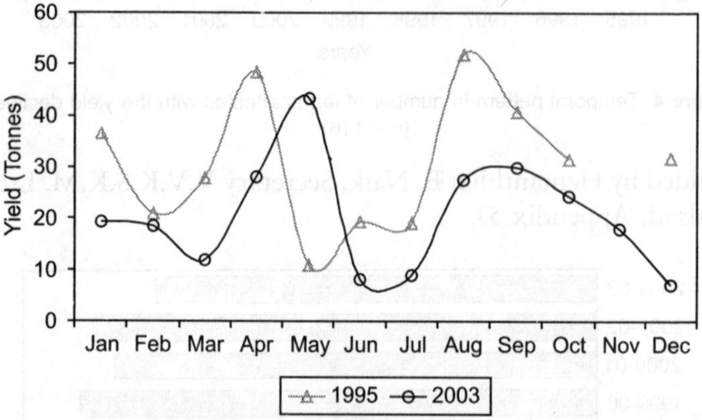

Figure 6: Change in the seasonal pattern in the yield of Sapota (at Saravali)

10. Factors Responsible for the Decline

There are many potential factors that could be hypothesised to be responsible for the observed decline in the productivity of plants (such as climate change, erratic rainfall, lack of sunlight penetration, pest incidences, changing water quality, dropping water tables, seawater intrusion, pollinator decline and emissions from the thermal power station). The global climate change has proven to be seriously impacting a variety of ecosystems (Bogya et al. 2000; McCarty 2001; Okazaki 2003) and has already been reported to have major impacts on agriculture in Europe (Ito and Oikawa 2002) and America (Peng and Apps 1999; Ni et al. 2001). The climate change has also been reported to effect changes in life history patterns of plants by changing the flowering pattern (Fitter and Fitter 2002). However many of these possibilities may could be ruled out in the context of Dahanu (Appendix 6), and some others need to be further investigated in detail using standardised methods before arriving at any comprehensive conclusion regarding their causative relation with the decline in Chicku yields. Following major factors important in the context of Dahanu could be identified for examination for the purpose of the present study.

1. Atmospheric pollution
2. Deterioration of soil quality
3. Decline in water quality
4. Decline of Natural enemies of pests
5. Pests and Diseases
6. pH change in soil and water
7. Pollination failure
8. Climatic changes
9. Low intensity of sunlight on plants (foggy weather)
10. Natural senescence of Orchard plants
11. Change in the farming practice
12. Decline in water table and saltwater intrusion
13. Nutrient deficiency

It may be noted that all these factors except a few, could be either directly or indirectly linked to the environmental pollution. However, quantitative assessment of the pollution load in the environment and its consequences on the ecological system of the area was not attempted during this study. The impacts of major possible factors and possible mitigatory measures are outlined below.

10.1. *High incidence of insect pests*

The pest incidence in the orchards has gone up significantly during the past few years. Apart from existing pest species crossing the economic threshold limits, new pest species have also emerged to compete with others. Presently there are quite a few serious insect pests in the orchards of Dahanu that are causing considerable economic loss (Table 4). According to the observations made by the farmers, earlier the bud borers (*Anarsia*

Table 4: Insects pests of Sapota prevalent in Dahanu orchards

Sl. No.	Pest Species	Target of attack	Season
1.	Bud Borer (*Anarsia achrasella*)	Flowers and buds	March-June
2.	Seed borer (*Trymalitis margarius*)	Young fruits	Oct-Jan
3.	Chicku Moth (*Nephopteryx eugraphella*)	Terminal leaves	Sep-Nov
4.	Fruitfly (*Bactrocera dorsalis*)	Mature fruits	-
5.	Mealy bug (Pseudococcid sp.)	Tender plant	-
6.	Stem borer (*Indarbela tetraonis*)	Bark	-

achrasella) was playing a beneficial role in their orchards by effectively pruning the excess flowers off the plants and thus helping in optimal fruit bearing of the plants. In the changed environment, they have changed their roles from a 'farmer friendly insect' to a 'destructive pest' and are eating away the whole flower bunches, possibly because of the population explosion of the pests because of the ecological stress in the system which has impaired the natural population control mechanisms of these pests probably due to the reduction in their natural enemy populations caused by agro chemicals and air pollution, and also possibly because of the reduced flower/bud availability for the pest populations.

Some of the latest varieties of Chicku plants developed by Agriculture universities are resistant to some of these pests [Eg: The Periyakulam variety (PKM-1) of Tamil Nadu Agricultural University, Coimbatore is resistant to bud borer)]. Farmers may be encouraged to cultivate such superior varieties in future as a long-term pest management strategy for minimising the losses.

10.2. *Age of the plants (Senescence)*

Examination of the data from orchards of various age classes indicated that the decline was not dependent on the age of the plants since the decline was reported in plants of all age groups. However, the degree of yield reduction showed a significant positive relation with age of the plants (Figure 7) indicating the higher susceptibility of the older plants.

10.3. *Indiscriminate pesticide usage*

Ever since the declining yields were felt in the farms, the farmers had been trying out various disaster mitigation steps.

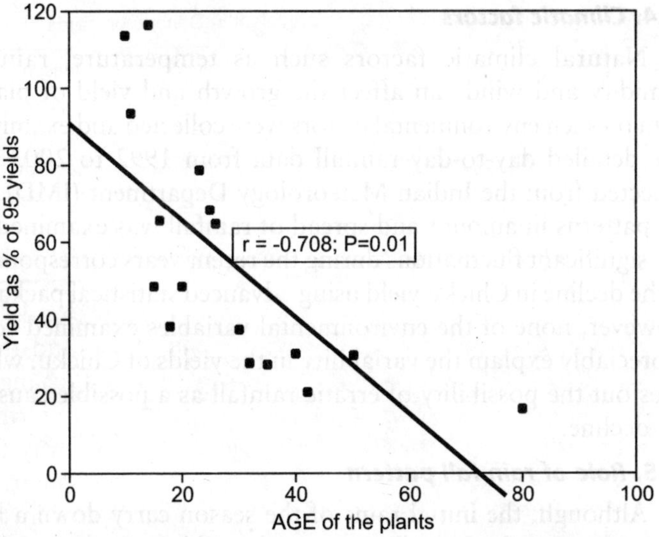

Figure 7: Age of plants vs yield (as % of 1995 yields)

The timely propaganda by agrochemical industries has been successful in weaning away some of the farmers from the traditional agricultural practices to the 'modern chemical farming', which has only resulted in worsening the situation by shifting towards more and more unsustainable and resource-intensive agricultural practices. Discussions with the farmers have revealed that many of them have resorted to indiscriminate and reckless use of even highly hazardous pesticides, which have resulted in the collapse of natural ecological balance that existed between the pests and their predators and parasites. Pesticides have wiped out more beneficial insects than pests. Widespread decline in the populations of red ants (*Oecophylla smaragdina*), social spiders and honeybees in the orchards that have adopted modern chemical farming testify the impact of agrochemicals and other pollutants on the beneficial fauna associated with the orchards.

There is an urgent need of generating awareness among the farmers on the hazards of pesticides and other agrochemicals which would help in guiding them back to the sustainable agricultural practices.

10.4. *Climatic factors*

Natural climatic factors such as temperature, rainfall, humidity and wind can affect the growth and yield of plants. Data on such environmental factors were collected and examined. The detailed day-to-day rainfall data from 1992 to 2003 was collected from the Indian Meteorology Department (IMD) and the patterns in amount and spread of rainfall was examined for any significant fluctuations during the recent years corresponding to the decline in Chicku yield using advanced statistical packages. However, none of the environmental variables examined could appreciably explain the variability in the yields of Chicku, which rules out the possibility of erratic rainfall as a possible cause of the decline.

10.5. *Role of rain fall pattern*

Although, the initial rains of the season carry down a high dose of atmospheric pollutants, and could prove harmful for the life forms. However, as the season progresses the pollution mitigation effect of the rain gets the upper hand, and it perhaps is the reason behind the better yields during the summer season (Refer Section 9.1). During the Monsoon season the impact of atmospheric pollutants on the plants are likely to be minimal compared to other seasons because of two major reasons.

(1) Lesser retention time of suspended particulate matter (SPM) and water-soluble gases in the atmosphere reduce the pollution load and

(2) The plants surfaces get thorough wash by the rains.

Because of low retention times, there is much less chances of direct injury to plants from the pollutants and there is lesser time available for chemical modification of the pollutants into secondary pollutants through atmospheric reactions. The cleaning up the leaf and flower surfaces by the rain greatly mitigates the impacts of SPM and acid depositions and help enhancing photosynthesis and pollination process. The loss of acidic depositions and other toxic pollutants from the soil through runoff also help the plant growth and restore the normal pH. This environmental improved conditions probably helped in better success rates in pollination, fruit setting and fruit retention during this season.

The development of flower into ripe fruit takes 8 to 10 months in Sapota. The monsoon season (June-August) flowers are responsible for the summer season yields of the next season and the summer season flowers are the source of harvest in the following winter. Hence the failure of the summer season flowers and success of the rainy season flowers strongly indicate the possible impact of higher pollution load during summer months on the plants.

10.6. *Decline in pollinator and predator populations*

There is a declining trend in the populations of beneficial insects in the orchards of the area. Bees and ants are becoming increasingly rare in the orchards. Apiculture has been affected severely and deserted bee boxes are a common sight in most of the farms. The predatory red ant (*Oecophylla smaragdina*) population also has been declining gradually from many of the orchards which probably helped producing conducive environment for better survival of insect pests, especially that of lepidopteran caterpillars. Dusting the trees with pesticides during the harvest period is practiced in certain farms and this could well be one of the causes for the red ant decline.

10.7. *Water—Availability, quality and irrigation methods*

Sapota requires regular irrigation for better yields. Chicku farms of Dahanu area are well irrigated and the majority of farms follows flood irrigation system, followed by drip irrigation and mixed irrigation systems (Figure 8). Although the water table has gone down during the years, there is no scarcity of water experienced in most of the farms. The yield from the flood-irrigated fields appears to be significantly better compared to that of drip irrigated fields (Appendix 4). This suggests the need for customising the drip irrigation to the needs of Sapota plants.

Among the water sources utilized for irrigation, bore wells were the source of water for majority of the farms (39.39%) while about a quarter of all the farms surveyed utilize open wells for irrigation. However the decline in the yield was recorded in

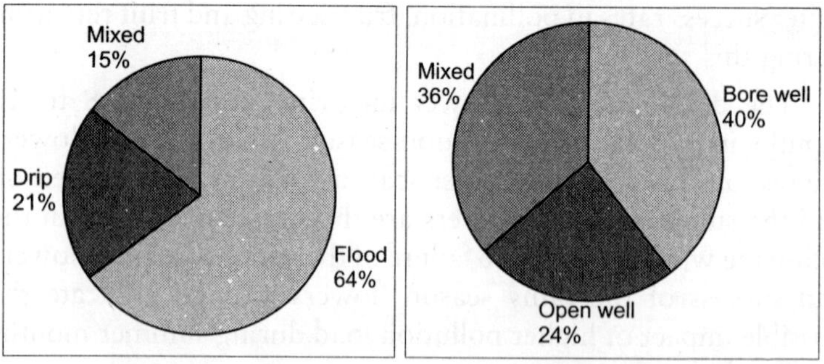

Figure 8: Irrigation method and Water source of the farms studied

all types of farms irrespective of water source and irrigation regimes (Figure 9). The result shows that the insignificance of the role of water sources and irrigation methods in the decline.

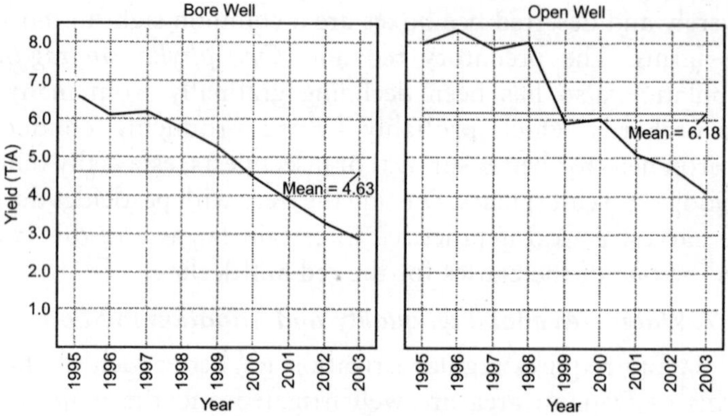

Figure 9: Comparison of production figures between farms with different water source

As none of the farms had reported water scarcity for irrigation, it appears to be a highly unlikely factor to cause the decline. Water quality was not tested during the present study and it need to be tested and monitored regularly for detecting any anomaly in quality.

10.8. *Nutrient deficiency*

Although the assessment of nutrient availability for the plants was beyond the scope of the present study, the field observations did not reveal any of the described deficiency symptoms (Nachegowda et al. 1992; Balakrishnan et al. 2000) among the plants. Hence the nutrient deficiency factor also appears unlikely.

11. Role of Industrial Pollution

The coal based Thermal Power Plant of Dahanu is the major large scale polluting industry present in Dahanu. Coal; being the most polluting fossil fuel currently in use, the coal powered thermal power plants are 'Pandora's boxes' of environmental hazards. The serious environmental hazards of coal based thermoelectric plants are well documented (Stuczynski et al. 1998; Negi and Meenkshi 1998; Mishra and Ramachandran 1998; Agrawal and Singh 2000; Baba et al. 2003; Mandal and Sengupta 2003).

However, adopting suitable fuels and pollution control measures it is possible to contain these impacts to a great extent, provided, such measures are implemented and operated meticulously. The negative correlation observed between the radial distance from DTPP and the mean decline at different places indicate a possible link between the yield decline and the pollution from DTPP (Figure 10).

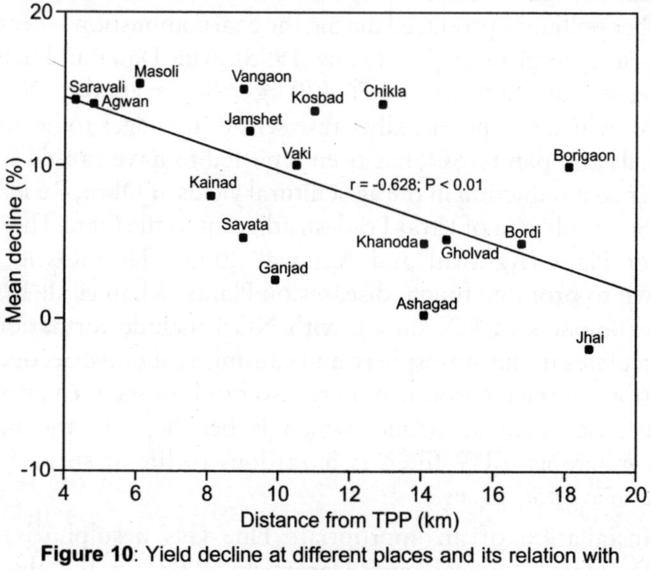

Figure 10: Yield decline at different places and its relation with distance from TPP

The 500 MW thermal power plant of Dahanu consumes about 8040 metric tonnes of coal each day and releases 76.4 tonnes of SO_2, 4.7 tonnes of SPM, and 72.35 tonnes of NOx every day (NEERI 1996) along with even more quantities of CO_2 and the ash slurries into the environment. Such emissions and effluents from the Thermal Power Plant contaminate air, soil and water and cause multifaceted impacts on the surrounding environment through multiple routes such as groundwater pollution (Davison and Lerner 2000; Adriano et al. 2002; Praharaj et al. 2002a; Baba et al. 2003), soil pollution (Singh and Narayan 1995; Lal 1996; Garcia Sanchez et al. 1999; Gulec et al. 2001), particulate matter pollution (Tucker 1998; Peden 2002; Fernandez et al. 2003), acidic depositions (McLeod et al. 1992; Mohren et al. 1992; Agrawal and Singh 2000; Brigden and Santillo 2002; Taylor and Frost 2004; Biyan and Yu 2004), radio active pollution (Papp et al. 2002) and so on. Following is a brief account of major pollutants emitted from the DTPP and their specific impacts.

The Sulphur Dioxide (SO_2): It is one of the major toxic gases released during the combustion of coal and is highly hazardous and cause negative impacts especially in the presence of fly ash—another pollutant produced during the coal combustion—because of their synergistic action (Tung 1995). The Dahanu Thermal power station emits around 76,400 kg of SO_2 every day (NEERI 1996), which can potentially cause serious damages to health of animals and plants. SO_2 has been reported to have caused 29 to 47 per cent reduction in the agricultural yields at Obra, Renekoot and Singrauli area of Uttar Pradesh adjacent to the Obra Thermal Power Plant (Agrawal and Agrawal 2000). The SO_2 is also known to promote fungal diseases on Plants (Khan et al. 1998). Other impacts of SO_2 (along with NO_2) include formation of particulates in the atmosphere and causing acid rains. Secondary reactions in the environment may also result in the formation of toxic gases such as ozone, which is beneficial in the upper atmosphere as a UV filter, is hazardous to life at surface and lower elevation zones.

Installation of an appropriate Flue Gas desulphurisation (FGD) plant (Miszcyk and Darowicki 2002) in the thermal power station can greatly reduce the SO_2 pollution. Latest alternate FGD technologies such as biological treatment systems are reportedly showing promising results compared traditional flue gas desulphurisation (Philip and Deshusses 2003).

Particulate matter: Of late the particulate matter pollution is being increasingly recognised as more dangerous than many of the gaseous pollutants with serious and irreversible repercussions on life forms, ecosystems and on structures and monuments (Mohanraj 2002). The dispersion and ground level concentrations of these factors are determined by a variety of factors such as physical characteristics of emissions, the stack, meteorological conditions, distance between source and receptors, condition of receptors, and so on.

Particulate matter emitted during the coal combustion is called the fly ash and is contaminated with various toxic and radio active elements (Negi and Meenkshi 1998; Sahu 1998; Zevenbergen et al. 1999; Flanders 1999; Papp et al. 2002; Gupta et al. 2002; Adriano et al. 2002; Praharaj et al. 2002b; Baba 2003; Gupta 2004). Apart from fly ash particles, the SOx and NOx gases also contribute to the load of particulate matter in the atmosphere by chemically transforming itself into acids and later into crystals of salts (sulphates and nitrates) in the environment.

Heavy metals: Heavy metals such as Cd, Cr, Cu, Ni, Pb, Zn and Hg present in the fly ash and bottom ash are toxic to plants, once become bio-available (Ariese et al. 2002; Gupta et al. 2002; Chaudhuri et al. 2003). Many of the heavy metals get enriched during the coal to ash conversion by combustion.

Trace elements: Many of the trace elements present in the Indian coal fly ash have deleterious impacts, often not explicit immediately after exposure. For instance the Selenium (Se), a toxic non-metallic element present in trace quantities in the coal get highly enriched in the fly ash (Sahu 1998). The accumulation of Se in the fish is reported to be creating a 'time bomb' situation in United States and elsewhere with substantial impacts on fish populations (Lemly 1999). The cascading impacts would be evident only when the time bomb explodes or the threshold selenium toxicity is exceeded and the lechates from the fly ash dumps and atmospheric fly ash originating from the Thermal power plant are capable of igniting the fuse of the time bomb by increasing waterborne concentrations of selenium and providing conditions favourable for bioaccumulation.

pH changes and bio-availability of trace elements: Changes in the hydrogen ion concentration (pH) is another important aspect of pollution from coal emissions. SO_2 and NO_2 emissions from the Thermal power plant emissions cause the formation of acidic depositions. It reduce the pH of soil and water and promote leaching of trace metals and results in enhanced bio-availability of these elements to the plants (Kanungo and Mohapatra 2000) which may prove toxic to the plants.

Other emissions: Other major emissions include greenhouse gases such as Carbon dioxide (CO_2), which cause rise in temperatures and associated impacts. Even the highly toxic and carcinogenic substances such as dioxins are also reported to be present in the exhausts of the coal fired power plants (Lima and Bachmann 2002; Hung and Lin 2003; Hylander et al. 2003).

12. Adverse Impacts of Pollution on the Plants

Ambient atmosphere plays a very important role in the development of plants. Nutrients absorbed from the soil form only a minute fraction of the biomass of plants and most of the biomass is produced photo-synthetically from the atmospheric gases in the presence of water. During this process, a large quantity of atmospheric gases passes through the plant tissues. Any noxious gas present in the atmosphere (such as SOx and NOx) and any obstruction to the gaseous exchange between the plant and its environment (Eg: stomatal closure by Particulate matter) can lead to undesirable and deleterious effects (Kayin et al. 1999; Agrawal and Singh 2000; Gupta et al. 2002). Serious impacts of thermal power plant emissions on the yields has been reported from the wheat fields around Obra thermal power plant (Sing et al. 1990).

Unlike many other Orchard trees, Sapota has the characteristic feature of flowering and fruiting continuously all round the year (Singh 1970; Davenport and ONeal 2000; Bari

2002). This makes the yield from Sapota plants more susceptible to stochastic environmental conditions round the year. Major deleterious impacts of by-products of coal combustion on the health of Chicku plants involves:

- *Impairment of photosynthesis*: Caused by fly ash depositions on the leaves and the presence of fly ash in the atmosphere in the form of particulate matter; both affecting the sunlight reaching the photosynthetic surface.
- *Interference with gaseous exchange and respiration*: Caused by deposition of fly ash particles on the leaf surfaces can chock up the stomatal openings of leaves.
- *Radioactive pollution*: The fly ash from Indian coal has high amounts of radioactive elements especially radon which can cause radioactive pollution and associated damage to the cells and genetic materials of the living organisms on continuous exposure (Papp et al. 2002).
- *Various negative impacts associated with pH changes*: Acids formed by the reaction of SOx and NOx with atmospheric water molecules cause acid rains and affect the health of the plants by causing physical injuries and affecting bio-availability of various nutrients (Table 5).

Table 5: Effect of acidity on plants

Effect on	Under Low pH (< 5.5)
Uptake of Nutrients	Retarded
Aluminium Toxicity	Develop due to free Al ion
Iron	-
Phosphate	Fixed
Calcium	Deficient
Availability of Anions, Mo & Bo	Reduced
Availability of Cations, Cu, Mn, Zn	Increased even to toxic levels
Certain Pathogens	Increased

Source: Wrigley 1981.

- *Pollination disruption*: Sapota is generally cross-pollinated and the pollination is effected by wind (Sulladmath and Reddy 1990; Bal 1997). Deposition of particulate matter on the stigma of the flowers cause drying up of stigma and act as a physical barrier for the pollens. Pollution also affects other pollinator populations such as bees, thrips and butterflies, which are also helpful for the pollination of flowers.
- *Withering of flowers and tender shoots*: Pollution induced acidic depositions and acid rains may cause charring symptoms on plants. Enhanced local temperatures caused by the higher levels of greenhouse gases such as CO_2 emitted from various contributing sources such as especially DTPP.
- *Promoting insect pests*: Caused by general weakness and lowered immunity levels caused by the environmental stress resulting from the pollution load in the environment. Apart from this, environmental pollution is likely to promote borer pests because of their specialized microhabitat preferences, which insulate them from the direct impact of the acid rains, toxic depositions and particulates to a great extent. Mild exposure to SO_2 is known to be conducive for fungal growth (Khan et al. 1998).

13. Evidences Indicating the Role of Pollution Factor

The emissions and discharges from the thermal power plant appears to be an important factor behind the decline due to the following reasons:

- The onset of yield decline from Sapota orchards coincides with the launching of Dahanu Thermal Power Plant (DTPP).
- The spatial pattern in decline shows a significant negative relationship with distance from DTPP, especially along the major wind directions (southwest and northeast). This indicates the role of emissions from the DTPP that are carried in the atmosphere along with the air movements.

- Changes in the pH of dew and rain and resultant loss in agricultural productivity indicate enhanced environmental levels of SOx and NOx which are among the major emissions from the DTPP.
- Enhanced atmospheric temperature has made the environment conducive for the proliferation of borer pests. High levels of greenhouse gas emissions from the DTPP would have also contributed to this.
- High rate of failure in the fruit bearing and development despite good anthesis can primarily be due to pollination failure and fungal attacks facilitated by acidic depositions on plant surfaces and other gaseous emissions from TPP.
- Higher impact on Chicku flowers during non-rainy seasons compared to rainy season also indicates the atmospheric pollution from DTPP. It may be noted here that the development of pollen and subsequent fertilization are highly sensitive to hydrogen ion concentration in the stigma and surroundings of the flowers.

14. Spatio-temporol Patterns in the Decline and DTPP

In order to examine the possible impact of pollution on the yield decline pattern, the Chicku yield data was examined with respect to the direction and distance of the farms from the Dahanu Thermal power plant. A comparison of the amount of yield reduction recorded from different directions from the TPP showed that the south-western parts are most affected followed by northeast (Figure 11). These are the two major wind directions during most time of the year in the area.

A comparison of the geographical patterns in the Chicku yields of 1995 and 2003 (Figure 12) indicates that the yield has drastically reduced in areas around the DTPP. It was further confirmed by the significant positive correlation observed between the present yield in percentage (as percentage of 1995 yields) and the distance from DTPP (Figure 13 and Table 6) indicating that the yield are better in farms further away from the Thermal Power plant. These results decisively indicate that the emission

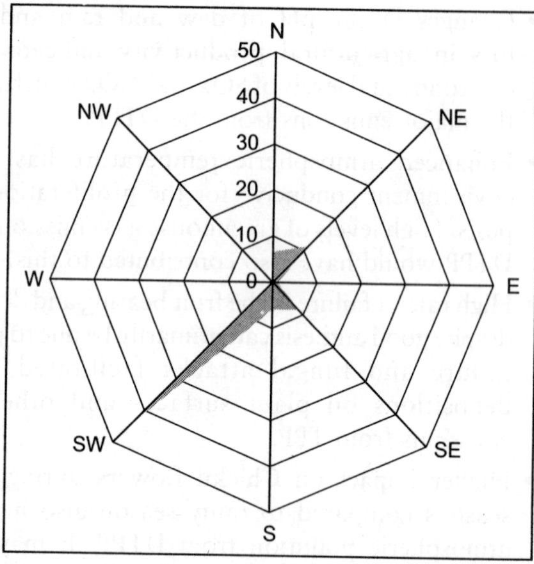

Figure 11: Directional pattern in the annual decline of Sapota yield from orchards

from the Coal fired thermal power plant of Dahanu has a major role in the decline of yields in the orchards.

Table 6: Correlation of Sapota yield with Distance from DTPP

		Distance from TPP	President yield in % of initial (fron data)
Distance from TPP	Pearson Correlation	1.000	.395*
	Sig. (2-tailed)		.012
	N	63	40
President yield in % of initial (fron data)	Pearson Correlation	395*	1.000
	Sig. (2-tailed)	.012	
	N	40	42

* Correlation is significant at the 0.05 level (2-tailed).

15. Estimated Losses

During the period of 1995 to 2003 the overall yield of Sapota from Dahanu's orchards dropped by around 60 per cent. Alarmingly, the trend in the yield has been taking a steady

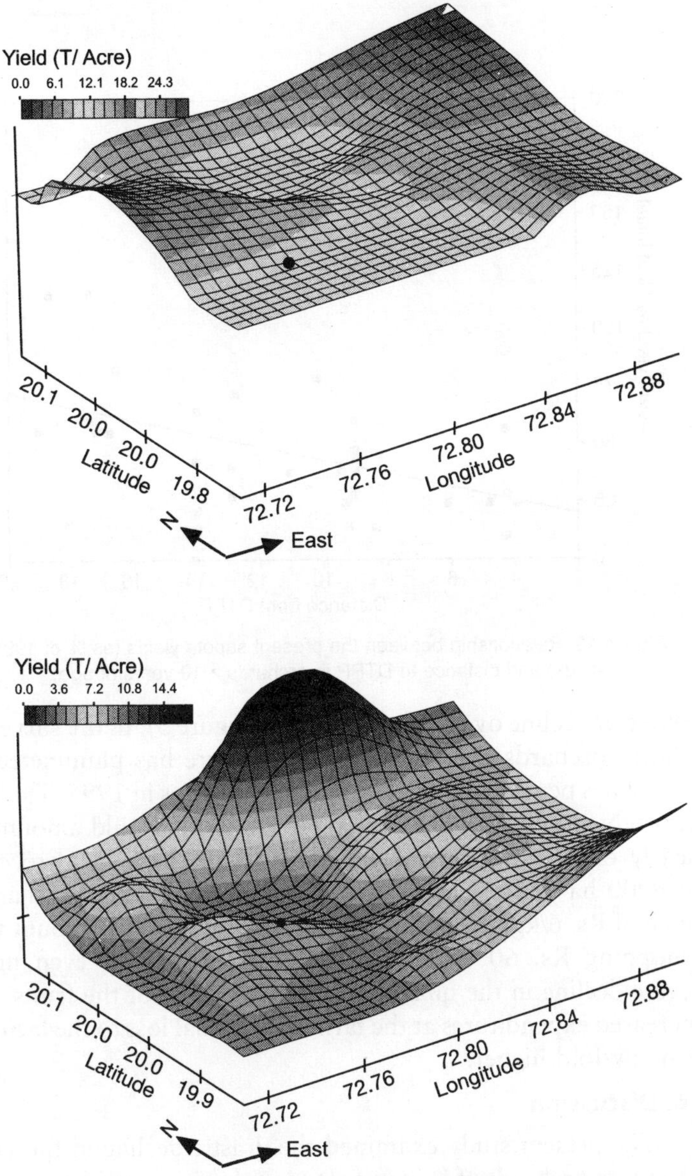

Figure 12: Spatial pattern in the yield of Sapota in Dahanu (Yield in southwest parts of the Taluka around Dahanu Thermal Power plant has seriously declined)

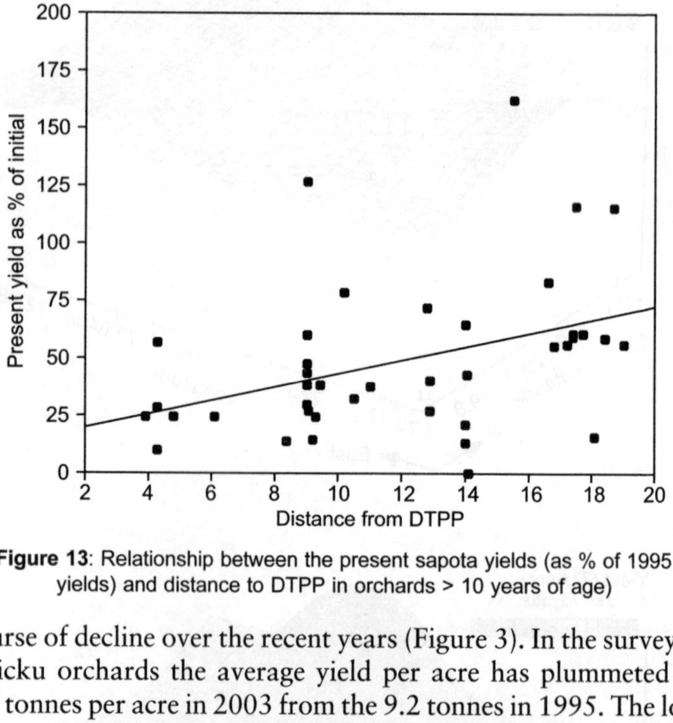

Figure 13: Relationship between the present sapota yields (as % of 1995 yields) and distance to DTPP in orchards > 10 years of age)

course of decline over the recent years (Figure 3). In the surveyed Chicku orchards the average yield per acre has plummeted to 3.7 tonnes per acre in 2003 from the 9.2 tonnes in 1995. The loss in production during the period (1995-2003) would amount to nearly one lakh tonnes if extrapolated to the official records of 6900 ha under Chicku cultivation. At an estimated market price of Rs. 6/kg, the monitory loss incurred would amount to a whopping Rs. 60 crores. Considering equally or even more severe decline in the quality and market value of the fruits and increased expenditures at the farms, the actual loss to the farmers is many-fold higher.

16. Discussion

The present study examined the drastic decline in the yield of Sapota orchards of Dahanu along with various environmental factors. Typically the decline was caused by failure during the fruit setting and later stages of fruit development. The flowering was quite satisfactory in most farms and it was the fruit setting and subsequent stages, which failed miserably. Apparently there are both obvious and unobvious reasons responsible for the decline. While pests and pathogens are the obvious factors, there are non-obvious factors such as pollinator decline, environmental pollution and resultant changes in the various biotic and abiotic factors.

The salient findings of the present study may be summarised as follows:

- The decline is evident among orchard plantations irrespective of age and variety of plants, water source, irrigation method, farming method (chemical/ biological). Although, the present study was focused only on the orchards, from opportunistic observations and incidental discussions with fishermen revealed that the fish and prawn yield from Dahanu creek and adjacent coastal areas has also suffered a major decline during the recent years (NEERI 1996).
- A point source of origin of the causative agent responsible for the catastrophic decline was evident from the severe decline in a limited area and a steady rise in the number of orchards experiencing the decline (from 1995 to 2003).
- The cause of decline is unlikely to be the climate change. In that case, the decline should have been felt in all the orchards more or less during the same time period and would have affected the orchards in a much larger spatial scale than the present case.
- Higher survival rate of the flowers and fruits during the time of least pollution load in the atmosphere (rainy season) and during the months of higher number of rainy days indicated the role of atmospheric pollution in the decline.
- Although the older plants are relatively more affected compared to younger plants, the senescence factor could not account for the yield reduction in relatively younger plantations.

- The indiscriminate use of pesticides and agrochemicals have caused serious decline in the natural predators of the pest species and other beneficial pollinator species, which has created conducive environmental conditions for the survival and multiplication of insect pests.

17. Conclusion

From the results of the present reconnaissance study, it may be stated that the pattern of decline indicates a causative relationship with the environmental pollution, especially atmospheric pollution and the consequent environmental impacts. The Dahanu thermal power plant is the single most likely source of this pollution and hence more stringent pollution control measures in DTTP especially to reduce the SO_2 and Ash emissions are imperative for the environmental health and long-term sustainability of Dahanu's orchard farms and other traditional livelihood supports.

The reckless usage of agrochemicals has worsened the situation by affecting the parasites and predators of the pests and thus promoting the pest populations in the orchards already under pollution stress. Appropriate campaigns for training the farmers in ecologically sound pest management strategies, minimising the usage of chemical pesticides and agrochemicals has to be undertaken by the concerned authorities. The sustainable benefits of biological farming as opposed to the short-term benefits and environmental hazards from of chemical farming need to be highlighted in such campaigns.

Dahanu notification (1991) prohibits any kind of polluting industry from being set up in Dahanu, in order to conserve the sensitive ecosystem of Dahanu from degradation. However, capitalising on the loopholes in the law, the huge coal-fired thermal power plant consuming large amounts of highly polluting coal continue to operate in the area, spewing out sizable quantities of pollutants into the environment. Operation of such a massive polluting industry in an ecologically fragile area flouting even the statutory environmental guidelines (NEERI 1996) is ironical and ecologically unreasonable. Moreover, considering the mounting scientific evidences highlighting the grave environmental consequences of coal combustion, it is unjustifiable to have a coal based thermoelectric station at an ecologically sensitive location, jeopardizing its sustainability. Hence, as an immediate mitigatory measure, the DTPP should drastically reduce the pollution using state of the art pollution controlling devices such as FGD. However in the long run, it should either have to switch over from coal to other environmentally safer and less polluting fuels such as Natural gas or be translocated to any other safer and ecologically less sensitive location elsewhere.

The present results are only indicative and further detailed and long-term studies need to be undertaken to monitor the air, water and soil quality parameters of the area in a regular 24-hour basis, in order to zero in on the exact nature source and environmental fate of the pollutants responsible for the catastrophe. Such a study is imperative to device long-term solutions and management strategies for the area as well. Only environmentally sound and comprehensive risk assessments based on scientific research followed up by prudent management actions can avert further disasters before it would be too late.

LITERATURE CITED

Adriano, D.C., Weber, J., Bolan, N.S., Paramasivam, S., Koo, B.J. and Sajwan, K.S. (2002). "Effects of High Rates of Coal Fly Ash on Soil, Turfgrass, and Groundwater Quality." *Water, Air and Soil Pollution.* 139 (1-4): 365-85.

Agrawal, M. and Singh, J. (2000). "Impact of Coal Power Plant Emission on the Foliar Elemental Concentrations in Plants in a Low Rainfall Tropical Region." *Environ. Monit. Assess.* 60 (3): 261-82.

Agrawal, S.B. and Agrawal, M. (2000). *Environmental Pollution and Plant Responses.* (Lewis Publishers).

Ariese, F., Swart, K., Morabito, R., Brunori, C., Balzamo, S., Slobodnik, J., Korenkova, E., Janos, P., Wildnerova, M., Hlavay, J., Polyak, K., Fodor, P. and Muntau, H. (2002). "Leaching Studies of Inorganic and Organic Compounds from Fly Ash." *Int. J. Environ. Anal. Chem.* 82 (11-12): 751-70.

Baba, A. (2003) "Geochemical Assessment of Environmental Effects of Ash from Yatagan (Mugla-Turkey) Thermal Power Plant." *Water, Air and Soil Pollution.* 144 (1): 3-18.

Baba, A., Kaya, A. and Birsoy, Y.K. (2003). "The Effect of Yatagan Thermal Power Plant (Mugla, Turkey) on the Quality of Surface and Groundwaters." *Water, Air and Soil Pollution.* 149 (1-4): 93-111.

Bal, J.S. (1997). "Sapota." *Fruit Growing.* Ed. J.S. Bal. 115-24 (Kalyani Publishers: New Delhi).

Balakrishnan, K., Rajendran, C. and Kulandaivelu, G. (2000). "Differential Responses of Iron, Magnesium, and Zinc Deficiency on Pigment Composition, Nutrient Content, and Photosynthetic Activity in Tropical fruit Crops." *Photosynthetica.* 38 (3): 477-79.

Bari, M. (2002). Chicku. 115.

Bari, M. and Bari, S.A. (2004). "Enironmental Study on Declination of Chicku Production in Coastal Parts of Thane and Navasari Districts." 13.

Biyan, Y.M. and Yu, S.W. (2004). "Forest Decline in Nanshan, China." *Forest Ecology and Management.* 51 (1-3).

Bogya, S., Marko, V. and Szinetar, C. (2000). "Effect of Pest Management Systems on Foliage- and Grass-dwelling Spider Communities in an Apple Orchard in Hungary." *Int. J. Pest. Manage.* 46 (4): 241-50.

Brigden, K. and Santillo, D. (2002). Hazardous emissions from Philippine Coal-fired Power plants.

Chaudhuri, D., Tripathy, S., Veeresh, H., Powell, M.A. and Hart, B.R. (2003). "Mobility and Bioavailability of Selected Heavy Metals in Coal Ash- and Sewage Sludge-amended Acid Soil." *Environ. Geol.* 44 (4): 419-32

Davenport, T.L. and ONeal, J.T. (2000). "Flowering and Fruit Set of Mamey Sapote [Calocarpum Sapota (Jacq.) Merr.] cv. Magana in South Florida." *Sci. Hort. Amsterdam.* 83 (1): 61-70.

Davison, R.M. and Lerner, D.N. (2000). "Evaluating Natural Attenuation of Groundwater Pollution from a Coal-carbonisation Plant: Developing a Local-scale Model Using MODFLOW, MODTMR and MT3D." *J. Chart. Inst. Water Env. Manage.* 14 (6): 419-26.

deBrito, E.S. and Narain, N. (2002). "Physical and Chemical Characteristics of Sapota Fruit at Different Stages of Maturation." *Pesquisa Agropecuaria Brasileira.* 37 (4): 567-72.

Dewan, R. and Chawla, M. (1999). "Of Development Amidst Fragility: A Societal and Environmental Perspective on Vadhavan Port." (Popular Prakashan: Mumbai).

Fernandez, A., Wendt, J.O.L., Wolski, N., Hein, K.R.G., Wang, S.J. and Witten, M.L. (2003). "Inhalation Health Effects of Fine Particles from the Co-combustion of Coal and Refuse Derived Fuel." *Chemosphere.* 51 (10): 1129-37.

Fitter, A.H. and Fitter, R.S.R. (2002). "Rapid Changes in Flowering Time in British Plants." *Science*. 296 (5573): 1689-91.

Flanders, P.J. (1999). "Identifying Fly Ash at a Distance from Fossil Fuel Power Stations." *Environ. Sci. Technol.* 33 (4): 528-32.

Ganjyal, G.M., Hanna, M.A. and Devadattam, D.S.K. (2003). "Processing of Sapota (Sapodilla): Drying." *Journal of Food Science*. 68 (2): 517-20

GarciaSanchez, A., Alastuey, A. and Querol, X. (1999). "Heavy Metal Adsorption by Different Minerals: Application to the Remediation of Polluted Soils." *Sci. Total Envir*. 242 (1-3): 179-88.

Gopal, M. and Gupta, A. (2001). "Has *Hirsutella Thompsonii* the Wherewithal to Counter Coconut Eriophyid Mite Scourge?" *Current Science*. 80, 831-36.

Gulec, N., Gunal, B. and Erler, A. (2001). "Assessment of Soil and Water Contamination Around an Ash-disposal Site: A Case Study from the Seyitomer Coal-fired Power Plant in Western Turkey." *Environ. Geol*. 40 (3): 331-44.

Gupta, D.K., Rai, U.N., Tripathi, R.D. and Inouhe, M. (2002). "Impacts of Fly-ash on Soil and Plant Responses." *J. Plant Res*. 115 (1122): 401-09.

Gupta, J.S. (1998). "Characteristics of Indian Coal Fly Ash and Its Utilization as Building Material." *Enironmental Impact of Coal Utilization*. 165-83.

Horvitz, C.C. and Schemske, D.W. (2002). "Effects of Plant Size, Leaf Herbivory, Local Competition and Fruit Production on Survival, Growth and Future Reproduction of a Neotropical Herb." *J. Ecol.* 90 (2): 279-90.

Hung, W.T. and Lin, C.F. (2003). "Use of Regenerated Ferric Oxide for CO Destruction and Suppressing Dioxin Formation in Flue Gas in a Pilot-scale Incinerator." *Chemosphere*. 53 (7): 727-35.

Hylander, L.D., Sollenberg, H. and Westas, H. (2003). "A Three-stage System to Remove Mercury and Dioxins in Flue Gases." *Sci. Total Envir*. 304 (1-3): 137-44.

Ito, A. and Oikawa, T. (2002). "A Simulation Model of the Carbon Cycle in Land Ecosystems (Sim-CYCLE): A Description Based on Dry-matter Production Theory and Plot-scale Validation." *Ecol. Model.* 151 (2-3): 143-76.

Kanungo, S.B. and Mohapatra, R. (2000). "Leaching Behavior o: Various Trace Metals in Aqueous Medium from Two Fly Ash Samples." *J. Environ. Qual.* 29 (1): 188-96.

Kayin, S., Tuncel, G. and Yurteri, C. (1999). "Local Air Quality Impacts Due to Downwash Around Thermal Power Plants: Numerical Simulations of the Effect of Building Orientation." *Environ. Monit. Assess.* 58 (1): 61-77.

Khan, M.R., Khan, M.W. and Pasha, M.J. (1998). "Effects of Sulfur Dioxide on the Development of Powdery Mildew of Cucumber." *Environ. Exp. Bot.* 40 (3): 265-73.

Lal, S. (1996). "Rehabilitation of Fly Ash Dump Yard of Shaktinagar Super Thermal Power Station Through Afforestation." *Indian Forester.* 122, 777-82.

Lemly, A.D. (1999). "Selenium Impacts on Fish: An Insidious Time Bomb." *Hum. Ecol. Risk Assessment.* 5 (6): 1139-51.

Lima, R. and Bachmann, R.T. (2002). "Pollutant Emissions from Modern Incinerators." *Int. J Environ. Pollution.* 18 (4); 336-45.

Mandal, A. and Sengupta, D. (2003). "Radioelemental Study of Kolaghat, Thermal Power Plant, West Bengal, India: Possible Environmental Hazards." *Environ. Geol.* 44 (2): 180-86.

McCarty, J.P. (2001). "Ecological Consequences of Recent Climate Change." *Conservation Biology.* 15 (2): 320-31.

McLeod, A.R., Shaw, P.J.A. and Holland, M.R. (1992). "The Liphook Forest Fumigation Project. Studies of Sulphur Dioxide and Ozone Effects on Coniferous Trees." *Forest Ecology and Management.* 51 (1-3): 121-27.

Mishra, U.C. and Ramachandran, T.V. (1998). "Environmental Impact of Coal Utilization for Electricity Generation." *Enironmental Impact of Coal Utilization.* 117-25.

Miszcyk, A. and Darowicki, K. (2002). "Reliability of Flue Gas Desulphurisation Installations—the Essential Condition of Efficient Air Pollution Control." *Pol. J. Environ. Stud.* 11 (3): 205-09.

Mohanraj, R. (2002). "Air Quality of Coimbatore with Emphasis on Respirable Particulate Matter." Ph.D. Thesis, Bharathiar Unievrsity Coimbatore.

Mohren, G.M.J., Jorritsma, I.T.M., Vermetten, A.W.M., Kropff, M.J., Smeets, W.L.M. and Tiktak, A. (1992). "Quantifying the Direct Effects of SO_2 and O_3 on Forest Growth." *Forest Ecology and Management.* 51 (1-3): 137-50.

Nachegowda, V., Sulladmath, U.V. and Vijayakumar, N. (1992). "Visual Nutrient Deficiency Symptoms In Sapota (Manilkara Achras Forsberg)." *Acta Hort.* 321: 566-73.

NEERI (1996). Examination of Regional Plan for Ecologically Fragile Dahanu Region and its Environmental Viability.

Negi, B.S. and Meenkshi, V. (1998). "Characterisation of Coal Fly Ash from Thermal Power Plants in India." *Enironmental Impact of Coal Utilization*. 143-64.

Ni, J., Zhang, X.S. and Scurlock, J.M.O. (2001). "Synthesis and Analysis of Biomass and Net Primary Productivity in Chinese Forests." *Ann. For. Sci.* 58 (4): 351-84.

Okazaki, K. (2003). "Various Combinations of Biological Pesticides and Agrochemicals—For Efficient Insect Pest Management in Apple Orchards." *J. Pestic. Sci.* 28 (2): 251-56.

Papp, Z., Dezso, Z. and Daroczy, S. (2002). "Significant Radioactive Contamination of Soil Around a Coal-fired Thermal Power Plant." *J. Environ. Radioact.* 59 (2): 191-205.

Parrotta, J.A., Francis, J.K. and Knowles, O.H. (2002). "Harvesting Intensity Affects Forest Structure and Composition in an Upland Amazonian Forest." *Forest. Ecol. Manage.* 169 (3): 243-55.

Peden, D.B. (2002). "Pollutants and Asthma: Role of Air Toxics." *Environ. Health Perspect.* 110: 565-68.

Peng, C.H. and Apps, M.J. (1999). "Modelling the Response of Net Primary Productivity (NPP) of Boreal Forest Ecosystems to Changes in Climate and Fire Disturbance Regimes." *Ecol. Model.* 122 (3): 175-93.

Philip, L. and Deshusses, M.A. (2003). "Sulfur Dioxide Treatment from Flue Gases Using a Biotrickling filter—Bioreactor System." *Environ. Sci. Technol.* 37 (9): 1978-82.

Praharaj, I., Swain, S.P., Powell, M.A., Hart, B.R. and Tripathy, S. (2002a). "Delineation of Groundwater Contamination Around an Ash Pond—Geochemical and GIS Approach." *Environ. Int.* 27 (8): 631-38.

Praharaj, T., Powell, M.A., Hart, B.R. and Tripathy, S. (2002b). "Leachability of Elements from Sub-bituminous Coal Fly Ash from India." *Environ. Int.* 27 (8): 609-15.

Rahmani, A.R., Muni, M., Ahmed, K. and Potia, U. (1999). *Biodiversity Documentation of Dahanu Taluka*. 1-108.

Sahu, K.C. (1998). "Coal and Fly Ash Problem." *Environmental Impact of Coal Utilization*.

Singh, J. and Narayan, D. (1995). "Changes in Soil Characteristics Around Coal-fired Power Plants." *Environment International*. 21 (1): 93-102.

Singh, J.S., Singh, K.P. and Agrawal, M. (1990). Environmental Degradation of Obra-Renekoot-Singrauli Area and its Impact on Natural and Derived Ecosystems, Project Report Submitted to the Ministry of Environment and Forests, Government of India, 14/167/84MAB/EN-2 RE.

Singh, R.D. (1970). "Studies on Flowering and Pollination in Sapota (*Achras sapota* L.)." *Indian J. Sci. & Indus.* 4, 65-72.

Stuczynski, T.I., McCarty, G.W. and Wright, R.J. (1998). "Impact of Coal Combustion Product Amendments on Soil Quality: I. Mobilization of Soil Organic Nitrogen." *Soil Sci.* 163 (12): 952-59.

Sulladmath, U.V. and Reddy, M.A. (1990). "Sapota." *Fruits: Tropical and Subtropical.* T.K. Bose and S.K. Mitra (Eds.). 565-91 (Naya Prokash: Calcutta).

Taylor, G. and Frost, D.L. (2004). "Impact of Gaseous Air Pollution on Leaf Growth of Hybrid Poplar." *Forest Ecology and Management.* 51 (1-3): 151-62.

Tucker, W.G. (1998). "Particulate Matter Sources, Emissions, and Control Options—USA." *Air Pollution in the 21st Century: Priority Issues and Policy.* 149-64 (Elsevier Science Pub: 1000 AE Amsterdam).

Tung, G. (1995). "Synergistic Effect of Fly Ash and SO_2 on Development of Cucumber (*Cucumis sativus* L.) Leaf Injury." *Enironmental Toxicology and Chemistry.* 14 (10): 1701-10.

World Bank Technical Paper Number 154 (1991). Environmental Assessment Source Book Volume III, Guidelines for Environmental Assessment of Energy and Industry Projects.

Zevenbergen, C., Bradley, J.P., VanReeuwijk, L.P., Shyam, A.K., Hjelmar, O. and Comans, R.N.J. (1999). "Clay formation and metal fixation During Weathering of Coal Fly Ash." *Environ. Sci. Technol.* 33 (19): 3405-09.

Appendix 1: Questionnaire used for the Survey

QUESTIONNAIRE (PAGE 1)

We are conducting a study on the horticultural practices of Dahanu.

Kindly help us by sparing a few minutes for filling up this questionnaire sincerely.

Name:	
Age Sex: M / F	Education: Nil/ <10th / 10th / Grad./ PG/
Place:	Panchayat:
Acreage of farm/field: .... ha	Sapota: .... ha
Major Crops	
Irrigation facility:	Yes / No
Water Source	Open Well / Bore well / River /
Irrigation Method	Open / Drip
Is their any change in flowering/fruiting patterns of your crops	
Is there any pest problem?	
If yes Since when?	
Any major change you adopted in your farming practice during the recent past?	
Of late is there any significant drop in production? If so since when?	Yes / No Since ....
Employment reduction	Mandays / year
What caused the decline?	Climate / Pest / Water / ....
Flowering/Fruiting pattern	Flower normal & Fruit failed
	Flower and fruit failed
Any suggestion to improve yields?	
Have you noticed any dust deposition on your plants?	Yes / No
Do you use any bio pesticide / fertilizers	
Are You a member of Krishak Samaj ?	Yes / No
If not, Why ?	
Would you like to get training on Organic Farming, fruit preservation technology etc.?	
Any other Information:	

FARM DETAILS (PAGE 2)

Name of the Farm :

Number of Bore wells :

Number of livestock : .... Cows; .... Goats; ....

Inter Cropping : Yes / No Bee-keeping ....

Crop	Variety	Age of farm	Employees: Regular	Employees: Harvest time	Fertilizer/ Pesticides Used	Irrigation type (Open/ Drip)	% Decline in yield	Any Other Remarks.

Chicku Production from your farm in	1995	1996	1997	1998	1999	2000	2001	2002	2003

Name & Address of Farm Owner:

Phone No: Place: Date:

Appendix 2: Location of farms surveyed

Farm Code #	Latitude (Degrees)	Longitude (Degrees)	Distance from TPP (km)	Angle to TPP (Degrees)
1	20.00	72.74	4.9	170
*2	-	-	15.5	184
3	20.11	72.76	17.2	186
4	20.11	72.77	17.4	187
5	20.00	72.74	4.9	172
7	20.11	72.76	17.5	187
11	19.99	72.76	4.3	208
13	19.95	72.83	8.3	272
15	19.99	72.75	3.9	190
18	20.06	72.74	11.3	178
*19	-	-	4.9	172
20	20.05	72.74	11.0	178
21	20.08	72.78	14.1	197
22	20.08	72.77	14.0	192
23	20.08	72.78	14.1	197
24	20.08	72.77	14.0	192
27	19.95	72.83	8.7	273
31	19.99	72.82	9.0	245
32	19.99	72.82	8.4	246
33	19.99	72.83	9.4	244
34	19.99	72.82	9.0	245
36	19.87	72.74	9.0	7
37	19.87	72.74	9.0	7
38	19.87	72.73	9.1	9
39	19.87	72.71	9.9	24
40	19.88	72.72	9.0	18
41	19.87	72.73	9.1	10
42	19.87	72.73	9.1	10
44	20.12	72.76	18.9	194
45	20.13	72.75	19.0	183
46	20.11	72.78	17.7	192
47	20.09	72.77	15.4	191
48	20.09	72.77	15.4	191
49	19.90	72.78	7.0	326

(*Contd...*)

Farm Code #	Latitude (Degrees)	Longitude (Degrees)	Distance from TPP (km)	Angle to TPP (Degrees)
51	19.87	72.71	9.8	22
52	20.00	72.83	10.2	244
54	20.08	72.74	14.4	178
55	19.94	72.88	14.2	277
58	20.00	72.75	5.0	187
59	19.94	72.87	12.8	277
62	19.99	72.79	6.1	227
63	19.94	72.79	5.1	-
64	19.94	72.78	4.1	291
65	20.04	72.75	-	185
66	19.87	72.74	-	0
67	20.00	72.74	4.8	180
68	20.09	72.74	14.7	179
70	20.09	72.74	15.2	178
*81	-	-	3.1	293
82	20.00	72.75	5.2	188
*84	-	-	9.7	-
85	20.09	72.73	18.7	181
86	20.10	72.75	16.6	181
87	20.11	72.74	16.6	181
*88	-	-	16.8	180
*89	-	-	16.6	181
90	19.94	72.79	5.2	294
93	20.02	72.78	8.1	204
96	20.12	72.76	18.4	185
*97	-	-	18.3	-
*99	-	-	17.4	174
100	20.09	72.74	14.8	177
109	20.10	72.76	16.3	186
110	20.00	72.75	4.6	185
113	20.10	72.75	16.5	182
114	-	-	9.6	-
*115	20.08	72.74	14.2	177
116	20.10	72.90	16.2	182

*Geographic position not recorded.

Appendix 3: Pattern of monthly rainfall at Dahanu during the last 12 years

Monthly Rainfall

Year	Jan.	Feb.	March	Apr.	May	June	July	Aug.	Sep.	Oct.	Nov.	Dec.
1992	-	-	-	-	-	414.8	349.9	703.0	405.4	19.6	-	-
1993	-	-	7.0	-	0.4	610.8	721.1	226.3	702.0	75.5	-	-
1994	26.2	-	-	-	-	845.9	780.6	632.8	161.1	7.9	0.3	-
1995	301.0	-	-	-	-	28.8	880.0	177.7	209.5	4.6	-	-
1996	1.4	-	-	-	-	213.5	1,062.4	271.6	137.7	871.0	1.4	-
1997	7.5	-	-	4.6	2.7	437.9	526.8	235.4	243.8	23.2	59.0	6.0
1998	-	0.2	-	-	-	504.3	504.0	419.4	677.4	278.4	0.7	-
1999	-	-	-	-	33.0	439.4	393.2	171.9	185.4	78.0	-	-
2000	-	-	-	-	94.9	81.3	843.9	404.9	28.2	20.2	-	1.2
2001	2.1	-	-	-	12.2	632.7	500.6	668.2	131.3	6.3	-	-
2002	-	-	3.5	-	3.8	1,020.7	85.2	578.4	31.0	0.2	0.2	-
2003	-	-	-	-	-	569.3	815.6	463.2	205.0	-	-	-

Number of Rainy days

Year	Jan.	Feb.	Mar.	Apr.	May	Jun.	Jul.	Aug.	Sep.	Oct.	Nov.	Dec.
1992	-	-	-	-	-	15	27	27	15	2	-	-
1993	1	-	1	-	1	15	27	23	20	8	-	-
1994	1	-	-	-	-	21	30	26	16	3	1	-
1995	1	-	-	-	-	8	27	21	11	5	-	-
1996	1	-	-	-	-	21	28	29	20	8	1	-
1997	2	-	-	1	1	21	26	29	19	-	5	2
1998	-	1	-	-	-	16	30	29	24	14	2	-
1999	-	1	-	-	6	15	20	26	19	7	-	-
2000	-	-	-	-	9	17	30	24	12	2	-	1
2001	1	-	-	-	6	23	31	29	14	5	-	-
2002	-	-	2	-	4	19	24	27	5	1	1	-
2003	-	-	-	-	-	20	30	29	26	-	-	-

Source: Indian Meteorology Department, Mumbai.

Appendix 4: Analysis of variance results showing significant variation in the yields between flood and drip irrigated fields

Anova

		Sum of Squares	df	Mean Square	F	Sig.
Mean yield (Tonnes/Acre)	Between Groups	436.009	2	218.004	11.785	.000
	Within Groups	684.463	37	18.499		
	Total	1120.472	39			

Appendix 5: Yearly detail of chicku pulling at Amalsad*

Year	Members	Mound	Amount	Average
1977-78	503	138328	5304405	29.74
1978-79	524	215137	5561285	25.85
1979-80	550	213759	6016168	28.31
1980-81	587	245989	8693711	35.34
1981-82	620	267373	9716253	36.34
1982-83	649	296705	11802914	39.78
1983-84	697	360110	11159808	30.99
1984-85	717	257272	11785914	45.81
1985-86	825	313501	14874120	47.44
1986-87	820	372925	19667959	52.74
1987-88	876	346190	25013149	72.25
1988-89	926	447963	24868456	55.51
1989-90	990	398831	25953304	65.07
1990-91	1037	536416	34366080	64.07
1991-92	1075	568269	43790030	77.06
1992-93	1136	619589	42525513	68.64
1993-94	1157	542956	40406785	74.42
1994-95	1321	656680	47359761	72.12
1995-96	1203	494523	49624133	100.35
1996-97	1339	626232	54519673	87.06
1997-98	1248	576874	64846844	112.41
1998-99	1224	537135	71320971	132.78
1999-00	1256	631841	61535858	97.39
2000-01	1244	587878	73361617	124.79
2001-02	1446	614642	67958557	110.55
2002-03	1459	539760	69820650	129.35

* Data kindly provided by Hemantbhai B. Naik, Secretary, V.V.K.S.K.M. Ltd., Amalsad

Appendix 6: Role of various factors in the declining yields

Sl. No.	Factors Examined	Results
1.	Pollution from the Thermal power plant	As the emissions from TPP contribute greatly to many of the possible problems examined. TPP is a grave concern and further studies should focus more on the issue.
2.	Soil quality	According to farmers, soil testing did not reveal any quality deterioration. May need further investigation.
3.	Water quality	Although the water table has gone down, however, water quality need to be monitored.
4.	Pollinators and Predators	Pollinator population is found to be greatly declined. Possibly by multiple factors such as acid rains and chemical pesticides. Beneficial insects such as Weaver ants (*Oecophylla smaragdina*), which were helping to keep the insect pests under check, have also reduced.
5.	Pests and diseases	This is a major problem. Borer pests and fungal diseases are prevalent. The deteriorating health and immunity of the plants has aggravated the problem. Mild Sulphur dioxide (SO_2) pollution is known to promote fungal growth.
6.	pH change	There are reports of acidic dew formations on the plants. Water and soil samples need to be monitored for pH fluctuations.
7.	Pollination barriers	Dust & Ash particle deposited on the leaves and flowers cause decreased photosynthesis and such depositions on the stigma of flowers act as an effective barrier for pollination, and thus result in reduction of yield from plants.
8.	Climate	Decline did not show any significant relation with the climatic factors such as rainfall. Higher number of rainy days favour.
9.	Changes in light regime	Because of the increased load of suspended particles in the atmosphere the light penetration through the atmosphere is hampered and it negatively affect the plants and other photosynthetic organisms.
10.	Senescence of plants	It is found to be not a factor, since farms in a wide range of age group has shown the

(*Contd...*)

Sl. No.	Factors Examined	Results
		decline. However the older plants are affected more (> 50 years).
11.	Farming practice	No major change in the farming practice in the recent past. Some farmers have started using pesticides and chemicals, but this was mostly adopted only after the decline was felt in their farms as a corrective measure.
12.	Water scarcity, decline in water table and saltwater intrusion	Although the water table has gone down slightly, presently there is no water scarcity in the Sapota farms of Dahanu and are mostly well irrigated. Hence, it is unlikely to be a factor. Saltwater intrusion is also not a major factor as many farms are located far from the coastline and the Sapota is known to be tolerant to salts in the soil (Sulladmath and Reddy 1990).

Appendix 7: List of butterflies recorded from Dahanu during the survey

Sl.No.	Common Name	Scientific Name	Family
1.	Brown Awl	*Badamia exclamationis* Fabricius	Hesperiidae
2.	Common Redeye	*Matapa aria* (Moore)	Hesperiidae
3.	Water Snow Flat	*Tagiades litigiosa* (Moschler)	Hesperiidae
4.	Angled Pierrot	*Caleta caleta* Hewitson	Lycaenidae
5.	Common Cerulean	*Jamides celeno* Cramer	Lycaenidae
6.	Common Pierrot	*Castalius rosimon* (Fabricius)	Lycaenidae
7.	Common Silverline	*Spindasis vulcans* Moore	Lycaenidae
8.	Dark Cerulean	*Jamides bochus* Cramer	Lycaenidae
9.	Pale Grass Blue	*Pseudozizeeria maha* (Kiollar)	Lycaenidae
10.	Pea Blue	*Lampides boeticus* (Lin.)	Lycaenidae
11.	Angled Castor	*Ariadne ariadne* Fruhstorfer	Nymphalidae
12.	Blue Pansy	*Precis orithya* Hubener	Nymphalidae
13.	Blue Tiger	*Tirumala limniace* Butler	Nymphalidae
14.	Chocolate Pansy	*Precis iphita* Cramer	Nymphalidae
15.	Common Baron	*Euthalia aconthea* Fruhtorfer	Nymphalidae
16.	Common Bushbrown	*Mycalesis perseus* Fabricius	Nymphalidae
17.	Common Crow	*Euploea core* Cramer	Nymphalidae

(Contd...)

Sl.No.	Common Name	Scientific Name	Family
18.	Common Eveningbrown	*Melanitis leda* Cramer	Nymphalidae
19.	Common Fivering	*Ypthima baldus* Fabricius	Nymphalidae
20.	Common Fourring	*Ypthima hubneri* Kirby	Nymphalidae
21.	Common Leopard	*Phalanta phalantha* Drury	Nymphalidae
22.	Common Map	*Cyrestis thyodamas* (Boisduval)	Nymphalidae
23.	Common Sailor	*Neptis hylas* Moore	Nymphalidae
24.	Danaid Eggfly	*Hypolimnas misippus* Linnaeus	Nymphalidae
25.	Great Eggfly	*Hypolimnas bolina* Linnaeus	Nymphalidae
26.	Grey Pansy	*Precis atlites* (Johanssen)	Nymphalidae
27.	Lemon Pansy	*Precis lemonias* Linnaeus	Nymphalidae
28.	Painted Lady	*Cynthia cardui* (Linn.)	Nymphalidae
29.	Peacock Pansy	*Precis almana* Linnaeus	Nymphalidae
30.	Plain Tiger	*Danaus chrysippus* Linnaeus	Nymphalidae
31.	Striped Tiger	*Danaus genutia* Cramer	Nymphalidae
32.	Tawny Coster	*Acraea violae* Horsfield	Nymphalidae
33.	Yellow Pansy	*Precis hierta* Evans	Nymphalidae
34.	Blue Mormon	*Papilio polymnestor* Cramer	Papilionidae
35	Common Mormon	*Princeps polytes* Cramer	Papilionidae
36.	Common Rose	*Pachliopta aristolochiae* Fabricius	Papilionidae
37.	Crimson Rose	*Pachliopta hector* Linnaeus	Papilionidae
38.	Lime Butterfly	*Princeps demoleus* Linnaeus	Papilionidae
39.	Red Helen	*Princeps helenus* Linnaeus	Papilionidae
40.	Common Albatross	*Appias albina* (C. & R. Fielder)	Pieridae
41.	Common Grass Yellow	*Eurema hecabe* Moore	Pieridae
42.	Common Jezebel	*Delias eucharis* Drury	Pieridae
43.	Common Wanderer	*Pareronia valeria* Fabricius	Pieridae
44.	Indian Cabbage White	*Pieris canidia* (Evans)	Pieridae
45.	Lemon Emigrant	*Catopsilia pomona* Fabricius	Pieridae
46.	Mottled Emigrant	*Catopsilia pyranthe* Linnaeus	Pieridae
47.	Pioneer	*Anaphaeis aurota* Fabricius	Pieridae
48.	Psyche	*Leptosia nina* Fabricius	Pieridae
49.	Small Salmon Arab	*Colotis calais* (Cramer)	Pieridae
50.	Yellow Orangetip	*Ixias pyrene* Butler	Pieridae

Planning and Implementation of Sustainable Development: A Case of Leather Industry in India

10

N.R. Vimala

Introduction

Indian leather industry has massive potential for employment, growth and exports. The industry exists in small, medium and large scale sector. There has been increasing emphasis on its planned development, aimed at optimum utilisation of available raw materials for maximising the returns, particularly from exports.

India is the largest livestock holding country with 220 million cattle, 90 million buffalo, 160 million goat and sheep population. There is 2.50 million workforce (30 per cent women). A source for 10 per cent global leather requirement with annual production value US $ 4 billion. It has enormous potential for future growth. The world import of leather and India's share is given in Table 1.

Table 1: World import of leather and India's share 2000-2004

Details	2000	2001	2002	2003	2004
World Import	14979.29	16208.67	15909.70	17532.66	18653.14
India's Export	381.49	459.25	508.83	555.71	577.00
% Share of India	2.55	2.83	3.20	3.17	3.09

Source: http:// www.leatherindia.org/statistics.asp

The leather and leather products are exported mainly to Germany, USA, Italy, UK and France.

The Estimated Production Capacities for Leather in the country is for hides (65 million pieces), for skins (170 million pieces) Leather Footwear (776 million pairs), Leather Shoe Uppers

(112 million pairs), Non-leather Footwear (960 million pairs), Leather Garments (018 million pieces), Leather Goods (060 million pieces), Industrial Gloves (052 million pairs), Saddlery (0.10 million pieces). Total No. of Units in Leather goods (55768), No. of Units in Organised sector (2477), No. of Units in Small Scale sector (53291).

The units involved in production of leather products are concentrated mainly in the state of Tamil Nadu, Delhi, West Bengal, Maharashtra, Uttar Pradesh, Punjab, Karnataka and Andhra Pradesh.

Institutional Mechanism

Various Ministries of the Government of India are involved in the sub-sector at the policy level. The Department of Industrial Policy and Promotion, Ministry of Industry is a nodal Ministry for formulation and implementation of policies for the development of leather and allied sectors. The UNDP assisted National Leather Development Program (NLDP) is executed by the Department. The Ministry of Commerce is responsible for formulation and implementation of policy for export of leather and leather products from India. The Ministry looks after the Council of Leather Exports. The Ministry of Environment and Forests is responsible for formulation of policy and implementation and promotion of environment protection projects. The Ministry through the concerned State Pollution Control Boards oversees establishment of pollutant discharge norms and their implementation.

Problem

As per the tenth five-year plan working group report for the leather sector the total number of tanneries in India are estimated to be 2091. Out of this only 1500 are functional. This has grown from 59 in 1950 to the present strength. Leather sector has been included among red categories of industries on account of the potential environmental impact of tannery wastes. Sustenance of tanneries is becoming increasingly difficult because of the alarming levels of environmental pollution caused by various tanning operations and practices. Complying with TDS norms has been found difficult by the tannery sector. Technology gaps with respect to cost effective treatment of total dissolved solids (TDS) and utilisation of some solid wastes have been identified. Tanning industry was confronted with serious environmental problems of tannery effluents. The tanneries are major agents of pollution of rivers and other water sources and are also annoying sources of public nuisance. This is due to the noxious chemical discharges, effluents and sludge emanating from tanneries, which are a constant source of danger to the environment. The importance of controlling pollution is generally not realised until considerable damage has already been done. Norms to discharge of treated effluents from leather industry were announced by central government called minimum national standards. However, the state governments were given authority to make norms more stringent taking into consideration the local conditions. The norms for minimum national standards are given in Table 2.

Table 2

Sl.No.	Parameter	Limits not to exceed
1.	pH	6.5-9.0
2.	BOD5-20°C	100 mg/litre(for surface waters 30 mg/litre)
3.	Total dissolved solid	100 mg/litre
4.	Sulphide	2 mg/litre
5.	Total chromium	2 mg/litre
6.	Oil and grease	10 mg/litre

Source: Sahasranaman A. "Tanneries in Tamil Nadu" Indian Leather, India, February 1997 Vol. 30/12, 63.

The crisis started with the closure of 400 tanneries in Tamil Nadu by the Apex Court in early 1996. A report prepared by the central leather research institute in 1993 states that there are 538 tanneries in the city of Calcutta in Tangra, Topsia and Tiljala clusters. In addition there are tanneries in Pagladanga and Karaya. The total processing capacity of all the tanneries is 850 tons per day. Average daily waste water discharge from Tangra area is estimated at 19,932 cubic meters, at Tiljala area 3,638 cubic meters and at Topsia 5,581 cubic meters. Thus, the detailed project report as stated approved by the Government of West Bengal, has proposed that the CETP should have a total capacity of 30,000 cubic meters per day for treatment of waste water from tanneries. It is planned to construct the CETP in

6 modules, each of 5,000 cubic meter per day capacity. Besides affording flexibility in meeting the demands of the tanneries in a graded manner, such an approach will also enable modifications and improvement in design in subsequent modules. It has been seen world-wide that the problems faced, in the operation and management of CETPs, vary from location to location and cluster to cluster. One of the important findings has revealed that for tanneries aggregated in clusters, the concept of the CETP may be the viable technical solution.

Some of the large scale tanneries have set up their own effluent treatment plants and many of these units have difficulty in meeting all the effluent discharge standards consistently in states like Tamil Nadu. The government has adopted several measures to deal with this problem from different aspects; supply of expertise, financial support, monitoring and enforcement.

In Calcutta, however, no facility for treatment of tannery effluent existed. In 1992 UNDP, New Delhi, had offered to assist the Government of West Bengal in setting up a CETP for one existing cluster of tanneries in Calcutta. The State Government however, had decided to relocate the tanneries from within the city to a new integrated leather complex (called Calcutta Leather Complex) at a location at Karaidanga on the Calcutta-Basanti Road, 16 kilometers east of Calcutta, in the district of South 24 Parganas. The reason for relocation was that the existing location of tanneries had come quite within the thickly populated areas of the city. A feasibility report had been prepared, for Calcutta Leather Complex (CLC), by the Central Leather Research Institute (CLRI), Chennai, under the UNDP assisted National Leather Development Program. (www.UNDP)

However, progress was restricted due to two reasons: (i) the government did not have the resources for building the new complex and (ii) a public interest litigation was filed in the Supreme Court of India on the subject of pollution caused by the Calcutta tanneries. The tanners assured the court that they would shift to the new complex. On the orders of the Supreme Court, National Environmental Engineering Research Institute

(NEERI), Nagpur and the West Bengal Pollution Control Board conducted a spot inspection and reported to the court that it was not feasible to put up CETPs at the existing location of tanneries and that the relocation of tanneries was absolutely unavoidable. Relocation has thus been confirmed as an unavoidable necessity.

In 1995, the Supreme Court had passed an order on the relocation of 598 tanneries located in the Tangra, Tiljala and Topsia areas of eastern Kolkata on pollution control grounds. The leather complex was conceived in 1995-96 as part of a Supreme Court directive against 500-odd highly polluting tanneries in the eastern fringes of Kolkata. Accordingly, a plan was laid to shift the tanneries in an integrated complex facilitated with common effluent treatment plants and other infrastructure facilities to avoid air and water pollution. However, the scope of the project was expanded towards attracting investment in leather and allied sectors from India and abroad. Out of a total of 1,100 acres about half was used for providing infrastructure including, roads, electricity, effluent treatment, processed water and other utility services. The rest of the space is used for industrial purposes. Of the same, about 320 acres is reserved for shifting the existing tanneries from Kolkata and new investments in the leather sector.

Overall, spaces were sold to 600 units so far including those to be relocated and new. Commissioning wise, the new entrants are the most active as merely six out of 20 units already set up are relocated from Kolkata.

While the tanneries at Topsia, accounting for 40 per cent of the production had capability to manufacture good products with forward integration, those at Tiljala were dominated by family-based units with small investments that turned out poor products.

In most cases obsolete technology was used. The disorganised nature of the industry and the absence of any umbrella organisation made any relocation difficult even after the State Government moved in to enforce the apex court order and appointed a private sector promoter to build the project.

Issues

In the Government of West Bengal, Department of Commerce and Industries and the Department of Cottage and Small Scale Industries are responsible for the development of this sector. Also the Department of Commerce and Industries has been entrusted with the task of planning and execution of the Calcutta Leather Complex project. M/s. M.L. Dalmiya & Co., Calcutta, has been selected as the Build-Operate-Transfer (BOT) contractor for the development of the CLC project.

The implementation of the sub program is the concern of the monitoring committee set up under the chairmanship of the Secretary, Commerce and Industries Department, Government of West Bengal, consisting of the SPC, representatives of the Ministry of Environment and Forests and the Ministry of Industry, Government of India, UNDP, CLRI, UNIDO, BOT and representatives of the tanners.

However, it is observed that though different groups of tanneries have their own associations, there is no single association of tanners to represent the interests of the entire tanning community in Calcutta. Hence, the Council of Leather Exports has taken up the role of co-ordination among the tanneries and is acting as an umbrella organization on behalf of the tanneries.

Relocating tanners face competition from tanners who were making fresh investment. It was reported that some of the tanners have till now failed to begin construction at the site as there was no funds support for the relocation and also their efforts to mortgage the land that they were acquiring drew a blank as two Government departments land revenues and industry were still squabbling over the issue.

West Bengal, which now has the third highest tanning capacity buys skins and hides from Bihar, Uttar Pradesh and Orissa. The tanners supply their products to New Delhi, Agra and Tamil Nadu, besides exporting. Although, more than 26 months have passed, very few tanneries have actually shut down. Many tanners chose to buy hides from outside and continue production since many of them belonged to backward classes and minority community and their family was dependent on this industry

Nine years should have been considered sufficient time for any major relocation project. But the Kolkata Leather Complex (KLC) has its own problems. Altogether 330 tanneries had been allocated land at the project site. State Industries Department sources reveal that about 15 tanneries have completed their construction at the site and were awaiting various clearances. Of these, nine were fresh investments, while only six had relocated from their original sites within the city.

However, the aim of the project is to create suitable infrastructure which can support and sustain this production activity within an environment friendly surroundings. At present, construction work at the project site is progressing in full steam. When fully developed, the complex will be the largest such complex in the world capable of processing 1000 tonnes of raw hide/skin per day.

Implications

Planning and implementation are very important factors for sustainable development. It is well recognized that tannery is a heavily polluting industry. Hence, in the earlier decades these were located at far of places away from the city and habitation. However, with the growth of the industries the population and habitation around the tanneries have proliferated in multifold. At the same time the legislations are not seriously implemented to prevent these habitation growth around polluting industry. Further, the groundwater around these areas get polluted and the habitants consume these polluted groundwater. As per the legislation tanneries are expected to have treatment plants and let the water out after treatment. There is a need for proper planning of location, habitation, effluent treatment so that shift of location at a later date with additional cost is avoided.

Wherever there is a problem of water shortage use of water should be minimum and also recycle wherever possible. The average consumption of water per kg of leather processed in tanneries in different states are listed in Table 3.

Table 3: Water consumption in leather processing (State-wise)

State	Average consumption of water/kg of hide (in litres)
Punjab	80-100
West Bengal	50-60
Uttar Pradesh	40-50
Tamil Nadu	22-30

Source: Tenth five-year Plan: S&T support system, leather sub sector", CSIR, India, March 2001.

In developed countries water is used at minimum level for leather processing. This helps in controlling pollution since effluents are reduced at the beginning. The quantity discharged is metered and taxed accordingly. The effluents are controlled at source.

In reference to compliance to norms in the State of Tamil Nadu Pollution Control Board has stipulated 30 BOD, 250 COD and 2100 ppm as TDS norms for surface discharge. CETPs and ETPs have been able to comply with BOD and COD norms but found difficult to reach TDS norms. In West Bengal CETP module is being designed for Calcutta Leather Complex complying with norms for sea discharge. The treated waters are planned to be discharged in nullahs bearing domestic sewer which is connected to an estuary and backwaters of sea. In Kanpur treated waste waters are admixed with domestic sewerage and diluted Norms relevant to domestic sewerage are applied.

Further in most of the developing countries pollutants discharge standards are rigid and overlook sight conditions. A gradual approach for installation of treatment facilities involve physco-chemical treatment followed by biological treatment and sludge handling. Instead of following the gradual method the tanner is put under pressure to install complete treatment system and meet all discharge limitations which is beyond his financial and technical means, particularly for small isolated units. Also, there is a need to educate tanners on environmental issues. Thus, environmental laws have been passed but implementation is not practical.

Conclusion

During the developmental phase environmental considerations seem to have been overlooked in the tanning sector though in individual cases safeguards have been built. Consideration of environmental economics of leather sector need to take into account several factors. Firstly, the loss of environmental and associated costs to the society living in the tannery cluster need to be quantified. Secondly, the option of closing of tanneries is associated with loss of opportunities and social loss. In addition, location of new units should be based on long-term land use planning to avoid relocation at a later date. Thirdly, though the pollution prevention and control form an area of one investment cost, providing zero environmental risk protection to neighbourhood is associated with large investment. Fourthly, the criteria for assessing environmental economic parameters should be sum of the investments needed for securing environmental protection of the community and its comparison with the social gain from the leather sector. Finally, it should take care of all costs of community service to be rendered by tanners and leather sector.

Tannery is one potential area for investment in India. The current industrial policy is very conducive to the promotion of the Indian leather industry both in terms of domestic trade as well as exports. Towards meeting the growing demand, India is aggressively looking at increasing its tannery activities and modernizing its facilities. Modernisation schemes introduced by Government of India is facilitating this activity.

REFERENCES

1. Bose, P.R. "SEZ Status for West Bengal Leather Complex", *The Hindu Business*, June 15, 2004.
2. Dutta, I. "Relocation Pangs of Kolkata Leather Complex", *Business Line*, June 22, 2004.
3. "Environment Program: Support Program on Common Effluent Treatment Plant, Calcutta Leather Complex", www UNDP.
4. Leather data, Indiandata.com
5. Sahasranaman, A. "Tanneries in Tamil Nadu", Indian Leather, India, February 1997, Vol. 30/12, 63-67.

6. "Tenth Five year Plan: S&T Support System, Leather Sub Sector", CSIR, India, March 2001.
7. "World Imports of Leather and Leather Products and India's Share 2000-2004", http:/www.leatherindia.org.asp.
8. Vimala, N.R. "Indian Leather Industry: A Perspective", *Yojana*, March 2003, Vol. 47/3, 45-48.

Women and Wildfire Management in Ghana: Learning from Feminine Principles

11

Agnes Atia Apusigah

Introduction

Wildfire management has become an important response to growing environmental degradation in Ghana. The destruction of the flora and fauna resulting from ravaging wildfires during the dry season, especially at the harmattan period has raised the concerns of not just environmentalists but also rural communities, non-governmental organizations as well as ruling governments and their agencies. Reaching its peak in the 1980s when the ravaging fires swept through the entire country and caused severe nationwide famine, various stakeholders have found preventive rather than curative options more appealing (Wasai, 2004; Tater, 2004; Millar 2004).

For those whose livelihoods are linked directly with nature, wildfires are a major factor for deepening poverty and social inequities. For those rural communities whose very ways of life are shaped by their livelihood activities in the forests and with nature, food crop farmers whose subsistence depends on crop yields and harvests and, rural women whose activities are organized around agriculture, the wildfires have meant dramatic life-changing decisions and intensified burdens. The burning of crop farms, economic trees, medicinal plants as well as animals which supply the foods, medicines and incomes of these peoples, has always meant the loss of livelihoods. That these categories of people remain the poorest in Ghana is not accidental (Ghana Government, 2003; GSS, 2002). The many disasters, including wildfires, that they have faced, natural and human, have wrecked grave havoc on individual and collectives lives with deep implications for communal and national development. The loss of livelihoods has meant increased poverty resulting in the slowing down of the pace of development at al levels. Hence, it has become a national imperative to find solutions to the unending destruction of lifestyles and livelihoods of these already vulnerable groups through wildfires.

Immediate responses to wildfires in Ghana have often taken the form of the mobilization of all forces to quench the fires and stop its spread. This has then been followed by various legislative and practical strategies aimed at finding lasting solutions to the yearly fires. Consequently, laws have passed such as the PNDC Law 49 of 1989 and Law 229 of 1992 (Tater, 2004; Millar, 2004; Ghana Government, 1990). After a long period of various negotiations and initiatives, in 2005, the Wildfire Bill was passed into policy (MLFM, 2005). Although, the new legislation has not come into full effect yet, it is yet one of many attempts that seek long lasting solutions to an endemic problem that affects vulnerable groups directly but also the nation at large since these groups provide the food and nutrition needs of the nation. Hence, wildfire management becomes not just a sectional but also national problem. It is not just a practical need but also a strategic interest. Its links with gender equality, food security, bio-diversity conservation, environmental degradation and poverty reduction place the problem high on the agenda of national development planning (Wasai, 2004; Ghana Government, 2003).

In earlier papers, I examine the broader question of wildfire management, its implications for various stakeholders and the lessons from their history of management (Apusigah, 2004, 2006). In this paper, I take up and interrogate the gender dimensions of wildfire management. I focus specifically on the stakes for women within the broader context of the continued exclusionary strategies adopted in wildfire management. Linking the problem of wildfires to the broader problem of environmental degradation, I show how patriarchal subsistent living and the paternalistic grand development projects have been pre-gendered. I argue that the devastating effects of wildfires and the high stakes for women necessitate the rethinking of ongoing management efforts. I argue also that the high incidence of wildfires, in spite of improved efforts, requires the initiation of new strategies that are appreciative of and incorporate all stakeholders, especially women at the grassroots, in the management process. In the process, I show that the patriarchal underpinnings of ongoing efforts deny the policies and resulting strategies the relevant insights for fostering effective collaborations. In that regard, I argue that by their problematic informings, ongoing efforts lack an appreciation of the gender implications of wildfires and as such fail to incorporate women's rich experiences in sustainable development to further the good cause of wildfire management. Attributing the situation to the loss of the feminine principle in development initiatives, I argue for a return to the feminine principle.

Drawing largely from the works of Vandana Shiva (1992, 1989), but also Maria Mies (1992, 1986), Valentine U. James (1995), Nel Noddings (1991) and Joan C. Tronto (1993), among others, I advocate for a return to the feminine principle for significant learnings. In the process I invite policy makers and program implementers to return to the feminine principle for critical learning in the form of the care, interdependence and ontological continuity necessary for effective collaborations in the management process while securing biological diversity and ecological balance. I conclude by arguing that for the sustained management of wildfires in Ghana, there is the need to engender the process even as effective collaborations are initiated and negotiated.

Wildfires and the Stakes for Rural Women

Feminists and gender workers in the area of development have and continue to challenge mainstream developmentalists and politicians on the inimical ways that development efforts are indifferent to women's concerns by minimalizing their contributions and experiences to inscribe and maintain dominance (Shiva, 1989; Boserup, 1970; Snyder & Tadesse, 1995; James, 1995). Couched in a paradigm that is heavily dependent on nature for survival and yet seeking to conquer rather than sustain nature's generous gifts, Ghana's development efforts,[1] framed in neo-liberal thinking, continue to violate the very persons in whose "interests" it purports to work (Nikoi, 1998; Tsikata, 2000). For women, the neo-liberal agenda of development continues to marginalise them as it fails to incorporate their concerns and rather alienates them from the so-called benefits of development. Thus, alienated, more and more rural women have to return to the land for sustenance and survival. In the particular case of wildfires, rural women are thrown into a desperate situation as their very survival is endangered. By their very positioning at the lower echelons of society and disadvantage in the socio-economic system, rural women suffer severe consequences during wildfire outbreaks. These include the loss of sustenance, loss of property, cause of physical harm and intensification of emotional stress.

Loss of Sustenance

In rural communities and among rural people, the bush or forest is the sole source of sustenance and survival. The people depend on forest resources to meet diverse needs. In addition to using the forests for the cultivation of food and cash crops, the natural forests serve as a key source for nature's generous supply of various food, fuel and cash items such as fruits, vegetables, wood, fibre, straw and meat. Women, especially those located in rural communities, also children, who occupy the lower echelons, depend on the forests for their survival especially during the off-farm season (dry season) when food stocks dwindle and rations reduce. During the early 1980s when Ghana faced its worse challenge of wildfire-related food shortages, women, especially those living in rural Ghana, had to live through very difficult times fending for their families.

Also, rural women depend on the bush or forests for supplementary income. In northern Ghana, especially among the Frafra people, women play major roles toward the cultivation of the land. Yet, they often have no control over the distribution of and proceeds from the sale of the products. As a result, these women cultivate their own crops on marginal lands to supplement allocated family food rations. They grow crops such as groundnuts, bambara beans, vegetables and potatoes, which they can call their own and which they can sell to support themselves. In addition, the bush becomes a major source of raising income. They pick straws for basketry, roofs and mats. Apart from personal use, women also sell or process them in order to add value and earn higher returns. As well, and very importantly medicinal, food and economic trees such as dawa dawa and shea nut, which have traditionally grown in the wild are major sources of sustenance for women. Women and their families have and continue to depend on these trees for cures, nutrition and income. The wildfires of the 1980s destroyed many such trees thus cutting down on a valuable source of sustenance.

Rural women also depend heavily on the forests and bushes for energy supply. From the cultivated fields they gather stalks, which are used for cooking and/or for sale. Stalks are highly combustible and as such hasten cooking while reducing the burden of long periods of fanning and cooking. Yet, the highly combustible nature of stalks often results in easy consumption by wildfires. This does not only put extra strain on women as they look for alternative energy sources, it also puts pressure on the forests, as they have to resort to the cutting of trees in an already degraded region.

Furthermore, the women of Ghana, in general and those in northern Ghana, in particular, depend on the forests and bushes for the treatment of several ailments. Apart from the fact that people living in rural areas have limited or no access to modern health facilities, cultural practices also prevent women from even accessing the limited facility. Studies reveal that the women, especially wives, have no control over their health needs (Hesse, 2000; Mohammed, 1999; Kerr, 1993). Their husbands, in-laws or heads of families determine access to health services. These studies reveal women delay or do not access existing health facilities because they are unable to obtain permission from the authorising party, who is more likely to be a mother-in-law, husband and/or head of family. Generally denied access to modern health facilities, rural women have had to depend on plant medicine. The leaves, roots, stems and bark of plants are used to cure a wide range of ailments. Ailments such as boils, typhoid, stomach-aches, malaria, diarrhoea, dysentery and fevers have been managed over the years through the use of herbal preparations. The use of herbal preparations, apart from serving the immediate purpose of disease management, also shields such women and their families from some of the side effects that are attributable to orthodox medical practices. The growing awareness of the potency and use of herbal medicine gives credence to the relevance of women's knowledges of herbs and their curative potential. Due to the importance of the forest in the lives of women and their families, any form of harm or destruction threatens their very survival women. Insofar as social relationships and interactions remain gendered with women located at the bottom, the forest remains a dependable ally that must be cared for and preserved.

Loss of Property

Another compelling factor that renders the stakes high for women in case of wildfires is the loss of property. Traditionally, denied ownership of critical capital resources for effective social participation, women, especially those in rural communities, have very little to show for wealth. They do not own the farmlands, the cattle ranches, household/compounds and/or hardly any of the cows, donkeys and sheep that form the resource base and wealth of rural communities. Often treated like the "property" of their male companions, married women in particular have restricted access to family property. As a result, whatever they acquire in the form of wealth is often done through personal efforts outside the family periphery. Even in some cases, women have no right to openly claim ownership of such hard earned property or wealth.

Among many northern Ghanaian communities, women are not encouraged to own certain properties that are considered male property. For instance, cattle are not considered suitable property for women. In cases where women acquire cattle on their own, they cannot openly declare ownership of such acquisition. Especially, when they have to sell some of their own acquisitions they have to do so through their husbands or other males. Consequently, women tend to invest in smaller and less-valued animals such as fowls, pigs and goats, in spite of all the risks. The sale of cattle is more cumbersome and often attracts some amount of scrutiny. Fictitious deals can easily leak out. The sale of fowls, pigs and goats are considered normal. Hence, one spouse can easily whisk away another's small animals and sell without question or detection. This stark lack of control over one's own resources pushes women to put their monies in less valued ventures.

Under the circumstance, women tend to invest their monies also on items over which they can have control. Without much money to start with, rural women tend to invest in crockery and/or clothing items. Their earnings are spent on buying pots, bowls, glass, chinaware and silverware, which are often trivialized. Such trivializing results in the acquisition glossed over in times of rescue. In cases where wildfires spread through communities and burned houses women are more likely to lose such wealth. Not viewed as the most valuable items of the house, rescuers often ignore such items. Even when the family is directly involved in the rescue mission, women would be required to help rescue what is considered valuable family property, which in reality belongs to male family members. As well, the natural tendency of women to put their needs last and others first, drives them to assist other family members before attending to themselves. Women will, therefore, often run to the rescue of children and the elderly in the family rather than take care of themselves. While the predisposition to help others first is laudable and must be encouraged, it is also important that the losses that women incur as a result are appreciated and acknowledged.

Physical Harm

During a workshop held in Bolgatanga for heads of decentralised departments and project/program desk officers, it came to light, during discussions that women were at high risk.[2] Several reasons were advanced to explain the high risk of women during wildfires. Participants enumerated some of the risk factors for women as follows:

- *During the performance of their roles*: During the dry season women have to walk various distances from the community in search for water, firewood and food, when their male counterparts will be at home. As bushfires occur during that time, they can be trapped.
- *The "inferior" physique of women*: It was argued that women compared to men were naturally weaker and slower. As a result during fire outbreak while men are able to run to safety women get trapped in the fires and have to be rescued.
- *Indecision on the part of women*: It was argued that women were hesitant about everything. They take time to make up their minds about things even during emergencies. During fire outbreaks, instead of running to safety women are caught in their indecision. They are often confused and have to be pulled out. If they are not rescued early they can be harmed.
- *Women are emotional*: When there is an outbreak of fire, women are often overwhelmed by fear. It was pointed out that women find the fury of bushfires very scary. They become too upset to move. Some even become unconscious and when they are not assisted they can be consumed by a wildfire.
- *Concern for family*: It was also argued that women will often try to alert other people around the vicinity of the fire and try to rescue them rather than run to safety first and call for help. They will try to rescue children, the elderly, sick and disabled, in the process they can endanger their own lives; and
- *Desire for property*: Another reason was that women are often hesitant to leave their properties behind while they run for their lives. In their hesitation and attempt to rescue the property they get trapped in the fires.

(Extracted from discussion notes, 2004)

From the above, it becomes clear that the socio-economic demands on and natural pre/dispositions of women put them at high risk in the event of wildfires. Also, in a conversation with personnel of the Upper East Regional Fire Department and National Disaster Management Organization the fact that women were of high risk was confirmed.[3] It is also important to note that in both instances (i.e., workshop and discussion fora) participants tended to blame women for risking their lives during emergencies. This victim-blaming tendency has serious implications for policy-making and programme planning, as perceptions tend to influence actions.

Emotional Stress

The cumulative effects of the loss of property, loss of sustenance and cause of physical harm on women is the increased emotional stress that women experience in the advent of wildfires. The loss and damages that are suffered when wildfires strike often translate in the increased pressure to find new sources to compensate for the loss. This limited sources for obtaining such alternatives often results in stressful conditions for women.

Additionally, as the key agents of family welfare, women are challenged emotionally in two ways. The first involves ensuring that family members are fed. Under normal circumstances women co-produce foodstuffs or staples with their male counterparts. Women's responsibility, however, goes beyond active cultivation to involve the processing for storage and the conversion of the stored food staples into meals for the entire family. Among most northern Ghanaian cultures, women have the added responsibility of providing food ingredients, fuel, turning raw staples to meal ingredients and ensuring that all family members were fed. This happens throughout the year requiring effective planning and management skills. In times of food shortage, women are faced with even harder challenges. In the midst of deprivation caused by losses to bushfires, in addition to the already tasking responsibility of the preparation and serving of food, women have to ensure that the food is available for preparation. When wildfires consume farms and destroy food crops creating shortage, women are presented with greater challenges resulting in increased stress.

A second and related issue arising from women's role as key agents of family welfare is the care that they are required to give to family members. Later discussions valorise this care-giving role. As caregivers of the sick, the elderly and children, women have additional challenges when bushfires strikes. Practical experience[4] shows that in addition to women and property, this category of family members is at a high risk of physical harm in case fire strikes. Being too weak or young to escape danger and/or too confused to make the right decision, the sick, the elderly, and children often fall victim when fire strikes the home. In situations like that not only do women endanger their own lives, as they struggle to rescue the more vulnerable, they also tend to blame themselves in case of endangerment. They take responsibility for not acting quickly to save the situation. As well, they have to nurse those injured to recovery by providing extra care beyond ordinary requirement. They have to nurse wounds of and provide special diets necessary for such sick and convalescing family members. In case of fatalities, women have to go through the pain of watching family members agonise to their death as they fight to keep them alive. These very stressful responsibilities take a toll on women's health and emotional balance.

The losses that women suffer and the stress that they experience serve to increase women's burdens in ways that are potentially life altering especially as they take up new roles in their homes and communities. The increased burdens and life altering roles that women are trapped in, in case of fire disasters, render the stakes too high for women. Consequently, it is expected that women would be treated as very important stakeholders and significant partners in matters regarding wildfires. Yet, as shown in the following section, ongoing efforts have and continue to show a lack of appreciation of this agency and tend to minimise and diminish women's contributions relative to the stakes.

The Problematic

While the above discussion depicts high stakes for women during the incidence of wildfires, ongoing strategic efforts continue to exclude them in the process. There is the stark lack of gender perspectives in the framing of policy and programming of interventions (MLFM, 2005; Apusigah, 2004, 2006; Ghana Government, 1990). This raises grave concerns necessitating interrogation. Anti-wildfire strategies do not reflect an appreciation of gender issues in the ways that the gendered structures work to favour males over females. The interventions hardly consider the different ways that wildfires affect females and males and as such hardly incorporate such understandings in management initiatives. Rather, strategists continue to rely on existing patriarchal structures and develop strategies that reinforce the unequal structures.

The strategies continue to treat women as perpetrators. Women are often viewed as likely to start fires from their domestic activities such as preparing stalks, preparing caustic soda and disposal of wastes from the fireplace. Yet, it is the case that men also set fires to roast various meats which they deny women from cooking. They conduct the burning of brush from compound farms. They also hunt for rats and bandacutts, which often results in the setting of fires to smoke such animals out of their abodes. On farms, women are suspected of not putting out fires set to prepare meals. Yet, it is the men who go hunting, smoke palm trees for palm wine, a local alcoholic drink and prepare land through slash and burn. Investigations for this paper revealed, among others, a viewing of women as key starters of wildfires and destroyers of the environment. Women were depicted as compromising ongoing efforts at the control and prevention of fires. It was indicated that women's careless handling of house fires such as the transferring of fire from one house to another, depositing of live coal in refuse damps, sending of children to fetch fires, burning of stalks for saltpetre, and burning of charcoal contribute to the starting of wildfires. Hence, it was argued that women must be educated on their negative roles. By extension, educational campaigns and gender programs seek to correct women. Although logical and to some extent true, I argue that programs that target and blame women for their negative roles are not likely to succeed, as implied in my analysis this far. Hence, I advocate for the re/channelling of wildfire management efforts toward drawing from the positive. It is for this cause, that I direct subsequent analysis.

Also, community level structures involve work with or by chiefs, elders, opinion leaders and youth who are often males. The community level actors are chosen on the basis of their sex and what is believed to be their sex roles and responsibilities. Traditional sex role differentiation defines women as weak and defenceless who must be located at home preferably in the kitchen and taken care of by men who have the strength and guts to defend. Guided by such conceptions, fire volunteer squads are normally the young men of the communities. These categories of people are often identified for their macho power but also for their role as decision-makers in communities. Since women remain excluded from these structures, their knowledges, skills and experiences are often also excluded in the development and implementation of strategies. In cases where women have come into the picture they have been targeted for education toward the stopping of their perceived negative practices which cause fires to start and go wild. Thus, women are often invited in already diagnosed and handed prescriptions for their ailments.

Even the fire department treats women officers as potential victims and not trained personnel with requisite knowledges and skills for fighting fire. Trained women in the Fire Department continue to work in secretarial, educational and dispensing positions. These women, who are trained fire fighters, are often assigned secondary duties as first aid dispensers rather than combatants. At fire scenes women can be found providing first aid to victims. The rest of the female personnel serve in traditional roles as clerks, typists and receptionists. They continue to play the traditional roles of caregivers. My concern here is not to dismiss this all-important ethic, as later discussions will show. Rather, I am concerned about the reductionism and essentialism of such thinking. Care giving is just one of the many qualities of women, which I will demonstrate in a later section is necessary for all humans. I am concerned when during outbreaks trained female fire fighters are hardly seen with fire tenders actively battling raging fires. At best, such trained female fire fighters are sent out to do community education, especially when they target women. During educational campaigns, female staff is sent to markets and women's groups to "talk to their fellow women," as was “slyly” put.[5] While this is strategic, in view of the rampant fire outbreaks in our market places and the possible ease in communication when females connect, to restrict trained female staff to such roles is problematic. It is problematic to the extent that it essentializes females and reduces their many capabilities to one, thus rendering such important and critical contributions discriminatory.

The enormous challenges that wildfires put on nation states, communities, families and individuals require that concerted efforts be made to harness and direct all relevant resources toward fighting back. As already noted, this will require dramatic shifts in the manner in which the policies and programs are framed. There is the need to move from thinking of wildfire management in masculinity terms such as control and prevention to include feminine principles that nurture relationships and collaborations. There is a lot to be learned from especially women in the management of wildfire. Their special relationship with nature equips them with knowledges and skills that I argue are potent for the strategic management of wildfires. These knowledges and skills have been developed through the nurturing of feminine principles.

Learning from Feminine Teachings

My investigations for this paper did not show any learning from women’s activities at the grassroots, which for me is an important pedagogical site with tremendous opportunities for effective wildfire management. At this site resides useful learning based on a principle that is rooted in countless years of experience of subjugation, subsistence and survival. These experiences of marginalization, especially at the grassroots, present challenges and realities that compel women to act in ways that enable them to appreciate existence differently. They have learned to collaborate and connect among themselves and with others in their contestations, confrontations and even embrace of marginalization, no matter how oddly. The feminine principle,[6] which guides women’s activities does not only enable them to cope but also presents important lessons for the whole of humanity.

The Feminine Principle

The feminine principle is that protective and nurturing quality and/or characteristic that enables women to maintain attachments to other humans, animals and even objects. It facilitates and furthers relationships and interactions between seekers (i.e., women/humans) and the sought (i.e., nature's bounty) in ways that are mutually respecting and supporting. It is rooted in the spirit of give and take rather than the extortion, exploitation and violence that attend patriarchal and paternalistic systems. At the grassroots, women's search for sustenance contributes to the preservation, regeneration, and interdependence rather than the destruction of the environment. Hence, an appreciation and embrace of the feminine principle will mean the careful guarding of one's actions and inactions and as well the initiation of activities that secure sustainable living. Such an appreciation will play a significant role in the re/channelling and shaping of the minimalist approaches to development planning and its violence antecedents along the path of multi-perspectivalism and non-violence. Contrasted with the masculine principle, which seeks to conquer and dominate nature, the feminine principle is couched in the philosophy of living peaceably with and the preservation of nature in all of its diversity. Although the feminine principle starts from and derives from women's experiences it is not essentially female. It is a quality that, I argue, is originally human, (i.e., exhibited in females and males), but which got lost during the patriarchal project of modernisation and industrialisation.

The loss of the feminine principle is traceable to the alienation that characterises modernity and its masculinisation of progress. The technical rationality that guides modernity's denying and abandoning of emotions, subjectivity and nature results in the mocking, devaluation and dismissing of the feminine. At the same time all things masculine were considered worthy, desirable and admirable. As the epitome of the denied and abandoned qualities, the feminine was rendered undesirable. This situation accounts for the negativism that is associated with the feminine and the dichotomised oppositionality that characterises gender relations. The disapproval of the feminine principle among

Ghanaians is exhibited in the persistent quest to avoid any feminine labelling. Ghanaian males, especially those of northern origins, whether young or old, work too hard to avoid being female. This has meant being tough and not showing emotions, not participating in activities considered feminine, not conversing with females, including wives, in public and even not using certain places (e.g., the kitchen) in the home. Thus denigrated, the feminine assumes an inferior position and becomes the object of subjugation and domination.

In light of the negativity that defines femininity and results from the oppositional dichotomy that characterises the patriarchal project, it is important to clarify the kind of femininity reclaimed. This requires the making of a distinction between femininity as an opposition to masculinity and femininity as a different way of being that is distinctively unique and valid. Femininity as an opposition to masculinity masks the very essence of the feminine; that is, peaceful co-existence. It also depicts the feminine as everything that is not masculine. Since the masculine has become the standard, the good, the feminine becomes substandard and evil/bad. A re/definition of the feminine as a unique and valid form of being is necessary for the expression of feminine difference in an empowering and liberating manner and not as one trapped in an oppositional binary. The distinction is meant to facilitate the process toward the abandonment of patriarchal mockery and devaluation of the feminine and all its negative connotations. It is also meant to rid femininity of all the negative tags, stereotypes and stigmas that devalue women and subscribe to subjugation and domination. A starting place will be the contestation and redefinition of the "gender division of labour in a society that devalues what women do" (Tronto, 1998: 354). In place, will be the establishment of a fair and just system where the different characteristics, roles and responsibilities of females and males are equally valued and rewarded and, social interactions and relations are based on the principles of equity. In the resulting system not only are women's activities but also what defines womanhood or femininity are appreciated and valued. By so doing, we can start from the different ways that women experience nature and learn from these ways.

It is also important to distinguish between the reclamation of the feminine principle as a means for protecting the remaining humane elements in humanity and the romanticization of nature as a pure, innocent and perfect state that must be reclaimed. The modernist loss of all things humane, fostered by the principle of the survival of the fittest rules, has meant violence to the not so fit. It has meant exclusion and even destruction of the vulnerable. To reverse this situation, it is important to reclaim our humanity by embracing the feminine. Yet, caution must be taken to avoid the re/romanticization of nature. The complexities that characterise nature make it at once appealing and repulsive. It is important, therefore, to keep these in view in the reclamation process. The feminine principle reaches beyond the romantic appeal where all nature is perfect, pure, and innocent and, the return is framed around an unquestioning embrace and preservation of it. Recognising the inadequacies of traditional systems and the problematic of an unquestioned embrace of nature, the feminine principle stays in constant contestation and confrontation even as it embraces and moves discourse out of the path of paralysis. Such actions are meant to force traditions to change elements that are outmoded and destructive. This calls for working with nature in a way that moves the relationship between humans and non-humans into a higher level of existence. This level of existence is meant to, in addition to being supportive of human existence, appreciate and work to maintain the existence of all life forms as well. At that level, the relationship is defined by care, independence and ontological continuity. These offer significant learning.

Significant Learning

I believe there are many lessons to be learned from women's experiences and relations to nature and with nature. The nurturing qualities of these relationships extend life and nature through the feminine guiding principles. Three critical ones are discussed here; namely, care, interdependence and ontological continuity.

Learning care

Post-modern thinkers, who have been concerned with questions regarding alienation, and of being, appeal to care as a means for the reclamation of the lost self (Crusius, 1995; Noddings, 1993). Contesting modernists grand theories and appeal to objectivity devoid of emotion resulting in the loss of being, post-modernists have sought to reclaim the self in all the ways that it influences and affects knowledge productions. As a result post-modernists compel a shift from the Cartesian rationality[7] that guides modernist thought to a new rationality that centres the subject in knowledge productions. In the works of Heidegger, Gadamar and Ricouer, on one hand and Gilligan, Noddings and Morgan, on the other, the connection between care and being is considered central to the understanding of human existence (Noddings, 1992; Crucius, 1991; Tronto, 1998). Heidegger views care as the Being of human life while Noddings sees it as a universal human characteristic (Noddings, 1992). The appreciation of the significance of care to human existence motivates the post-modern call for the return to the ethic of care (Noddings, 1992; Tronto, 1998).

In calling for the return to the care ethic, Gilligan, Noddings and Tronto, make a distinction between feminine and masculine modelling of care. These distinctions result in two types of care: 'care of' and 'care for'. 'Care of' is general and almost detached. This modelling of caring, typical of the masculine ethic, can be provided from a distance. On the other hand, 'care for' is personal and relational. The feminine ethic of care, which is built on a close relationship between the 'carer' and the 'cared', an attachment or dyad is formed and maintained. In caring for others and things, women are engrossed in a relationship of appreciation, concern and giving. Both the caregiver and the recipient of care recognise and appreciate their different investments in the process. This mutual feeling of appreciation sustains the relationship and fosters the continuance of care.

Women as caregivers play significant roles in the nurturing of not just household members but also the resources that are used to provide such care. Care as a feminine ethic underlies women's concern for others and the readiness with which women express their emotion. More often than not the emotionalism that women exhibit in times of crisis and need is used as a weapon to deny social participation and dismiss their contributions to the socio-economic system. Yet, this ethic is very essential in the maintenance of life. Indeed, the ways in which women's subsistent activities foster the maintenance of life forms and the environment has been attributed to this ethic (Shiva, 1995, 1989; Escobar, 1995; James, 1995; Dankelman and Davidson, 1988). Women's activities at the margins of agriculture and in the socio-economic system enable them to stay in close touch with important yet often dismissed elements of the biologically diverse world. The need to conserve the limited resources available to them at the margins requires that all efforts be made to extend the supply of such resources. Hence, women have learned to harness and harvest such resources in ways that extend life and supply. This learning is reflected in for instance the way that women harvest young leaves leaving the tender ones to support the life of the plant as well as to mature for future use while over matured leaves are left to fall out to complete the life cycle of the plant and to nourish the roots of the plant.

An appeal to the care ethic will require that plants, on which humans depend so much, be viewed as part of the cycle of life. Human dependence on plants for food, energy, medicinal and/or shelter, require a viewing that can reflect plant life as an extension of the cycle of human existence. To foster this relationship, humans, who tend to possess the destructive edge, will have to take steps to ensure that plant life is sustained. This calls for effective land management initiatives of sort not available or at best low key in Ghana. Effective land management will entail shifting from the traditional notion of using land efficiently to support human needs irrespective of the implications of such use for other beings, to embrace the culture of conservation more seriously. In that case, it will mean not burning the bush out of mischief but doing controlled burning because one appreciates the effect of wildfires. For instance, non-violent means and ecologically sound measures could be used for hunting. It will also mean that in burning there is the recognition of participating effectively in the process of completing the cycle of life rather than disrupting the effective completion of that cycle. This will mean burning dead wood, which can be considered as recycling and controlled slash and burn so that new life can begin.

The kind of effective land management and active conservation availed in feminine teachings require not just legislation and enforcement but an understanding of the dyad between humans and non-humans in ways that result in people acting voluntarily to maintain the link out of a position of care. If there is mutual and collective understanding there will be no need for legislation or enforcement at all. There is a lot to learn from this fundamental yet critical principle of sustainable living. When there is the recognition of the criticalness of maintaining the relations and securing the lives of all those with vested interests in it, there will be room for interdependent rather than dependent living.

Learning interdependence

The feminine principle also teaches interdependence, a crucial ingredient for fostering and sustaining necessary and effective collaborations among stakeholders toward the sustainability of bio-diversity. At one level, interdependence offers all stakeholders an opportunity to recognise and appreciate one another in the unique individual and collective ways that they are located within the life cycle and the need to keep their energies flowing through in a mutually reinforcing manner. The sustenance of the life cycle becomes at once individual and collective responsibility in the race against annihilation of not just sections but all components of the cycle. Hence, the weakening of any part or component of the chain becomes a matter of concern for all. At another level, an extension of the other, all components, viewing themselves as stakeholders, find a reflection of themselves in all others. Such a reflection moves them, individually and collectively, to work toward the enhancement of lives of their own kind and those of others within the cycle. It enables them to see beyond the differences in life patterns and experiences to an appreciation of a common fate that is tied through the dependence on one another. This has been what women's movements all over the globe have been about.

Locally and globally, women have organised among themselves in the struggle against annihilation. Women have organised to contest oppression, domination and subjugation. The struggles against patriarchal social relationships and paternalistic projects that have and continue to victimise women have gone on in spite of the myriad differences among women world-wide. Women and feminist contestations have only not targeted mainstream patriarchal and paternalistic systems but also the undue generalisationn and/or particularisation of women's experiences. Women have also organised in contestation and confrontation of things local and global. Women's organising of any form or creed seeks and maintains the connections among themselves and with others that share in their vulnerability. Women have mobilised to fight for their rights as well as the rights of their children and even male family members. They have organised around issues of human development and ecological balance. The experiences of women's collective actions reflected in the work of the Sontaa Nontaa Women of Ghana, Chipko Women of India, Seikatsu Club of Japan, Liberia Refugee Women in Ghana, Annang and Ogoni Women in Nigeria as well as the Green Belt Women of Kenya have been aimed at mobilising various categories of women and actually embarking on programs and projects in defence of their annihilation and survival. As well, activists such as Stella Awumbila, Vandana Shiva, Mehda Patker, Lois Gibbs, and Wangari Mathai, to name just a few, have struggled individually as well as mobilised other women and even communities in struggles for their right to life. These causes are wide ranging including social, economic, environmental and cultural. Rights advocacy has also included human, religious, social, gender, child, civil, environmental, and cultural rights. The challenges that such organising pose for the mainstream patriarchal and paternalistic project, are demonstrative of women's appreciation of the complexity of social relationships and interactions, and of change and diversity.

In recent decades, women's organising has become more and more sophisticated. While this sophistication is necessary for contesting the very complex apparatuses of subjugation and domination, it is also important to appreciate the humble roots of such complex processes. Persistently excluded and left with limited resources on which to function, women, from the beginnings of time, have relied and depended on one another for all kinds of support, emotionally and physically, locally and globally and, internally and externally. As a result women have learnt to and actually appreciated the very meaning of co-existence and the principles of interdependence.

Interdependence among life forms demands that there is peaceful co-existence and mutual support. This requires that all parties take active rather than passive roles in the struggle the survival of all. It also requires persistent and continuous networking. This, women the world over, have done very well. In spite of growing resistance and backlash, women maintain their stance of confronting and contesting dominance. They continue to fight and talk back using varied means and resources. Perhaps efforts at fire fighting can tap into these very basic principles of mutuality and collaborative endeavours to initiate and negotiate the process more effectively if live is to continue.

Learning ontological continuity

Central to the feminine principle is the maintenance of ontological continuity. The very existence and being of human kind is dependent on this principle. The drive to survive and dominate the world makes ontological continuity an imperative. From simple traditional to sophisticated modern societies ontological continuity has been sought. Ancient societies, like traditional societies, sought to maintain and reproduce themselves through the development of technologies that enable them to not only cope in their environment but also defend themselves against foes. This process has been maintained and sustained by modern societies. The search for survival and maintenance of human existence has lead to the kinds of modernisation processes that have proven inimical to life. Revolutions in science and technology have led to transformations in human existence. While such revolutions have been meant to foster progress in human existence and improve in the quality of life, it is also the case that they have also been marked by alarming levels of violence (Alvares, 1994; Shiva, 1993; Mies, 1993). As already noted, the greedy search to conquer nature have thrust humanity into the environmental degradations that have compounded the challenges of rural folks and compelled them to engage in practices that create the problem of bushfires. Above all, these revolutions have not generated any permanency in human existence. The human species, like all other forms of existence, continue to occupy a temporal place in the world. Temporality remains inevitable in spite of scientific and technological progress. This inevitability demands that we seek alternative and complementary ways of understanding ontological continuity. The teachings of the feminine principle provide possibility for learning such ways. At the risk of essentializing, I call for a return to, an appreciation of and learning from the process of reproduction as experienced by women. Although reproduction is not the preserve of women, the ways that women live through and manage their roles in the process has significant teachings for humankind.

The reproduction process is central to human existence. It is this role which fosters the survival and sustainability of humankind kind. In reproduction lies the difficult and challenging task of procreation. Through reproduction humanity manages temporality in order to sustain some level of permanency. As mortals, humans have terminal life span, however, through reproduction humans have remained on the face of the earth since their arrival. Humans maintain permanence through re-generating themselves. In the re-generation process lies learning that can inform bushfire management processes. While such learning is central to the entire process the unique ways that women participate in the process is where significant learning is located. Such learning is often lost in the process of development or rather mal-development. As argued by Shiva (1989) and implied by Mies (1986) patriarchal development has resulted in the death of the feminine principle, which needs to be reclaimed. Shiva sees the reclamation in not only women but also males and natures. She points out:

> The recovery of the feminine principle is based on inclusiveness. It is a recovery of nature, woman and man of creative forms of being and perceiving. In nature it implies seeing nature as a live organism. In woman it implies seeing

> women as productive and active. Finally, in men the recovery of the feminine principle implies a relocation of action and activity to create life enhancing, not life reducing and life-threatening societies. (Shiva, 1989: 53)

In Shiva's conception, a drastic revolution in the viewing of nature, women, and men is required. The recovery process is meant to engender mutual respect and collective action. It also requires that all actions, and/or even in actions, that are destructive such as bushfires are abrogated. Indeed, it makes sense to advocate for the recovery of the feminine principle in these times that bushfires that ravage lands and communities have become the norm rather than the exception. This recovery is necessary because of what has been called the death of the feminine principle in patriarchal and paternalistic quest for progress. For Shiva, women, nature and men are all victims of this process. As she points out:

> The death of the feminine principle in women and nature takes place through the association of the category of passivity with the feminine. The death of the feminine principle in men takes place by a shift in the concept of power from empowerment to domination. Self-generated, non-violent, creative activity as the feminine principle dies simultaneously in women, men and nature when violence and aggression become the masculine model of activity, and women and nature are turned into passive objects of violence. (*ibid.*)

The misconstructions that characterise the process are evident in the ways that progress is framed often as modernization and industrialization. The process itself is often marked with increasing rates of vulnerability and exclusionism as reflected in the poverty situation in Ghana. Indeed, Ghana's development process, under Economic Recovery Programme and Structural Adjustment, Vision 2020 and now HIPIC (i.e., Highly Indebted Poor Countries Initiative) are reflective of the violence that characterises progress. Promises of non-violence remain on paper and at best nominal. As well, in the fight against bushfires can be found reminisces of domination. The exclusion of women's roles and their concerns and the male-centric language pointed to earlier reflect violence rather than non-violence.

Another way by which the death of the feminine principle becomes evident is in the framing and valuing of female and male roles. As development has progressed, female work has been considered as reproductive while that of males has been considered productive. An explanation by the Canadian Council for International Co-operation (1991: 16-17) gives meaning and context to this role differentiation as follows:

> Every society assigns different tasks to men and women (often called the sexual division of labour). In most cultures, both women and men do productive work (produce goods or services for income or subsistence), although in most specific situations, productive work is clearly divided into men's tasks (ploughing, working with machinery) and women's (planting, weeding, teaching children). Women are mainly responsible for reproductive work (e.g., cooking, washing, cleaning, nursing, looking after children, building and maintaining shelter), to reproduce and maintain the labour force.

This gendered definition of labour and the specific responsibility that is often assigned to women in reproduction has often been misconstrued. The misconstruction arises when the productive aspects of women's work is glossed over and lumped into reproduction. Often arguing that men's work is labour intensive women's productive work is supportive of men's labour, women's contribution to production get minimised and dismissed as no work. The end result is that both women's productive and reproductive work is categorised as supportive and considered welfare roles, which must go unappreciated and unrewarded. A difficulty and indeed the challenge for women is the minimisation and diminishing that these equally important contributions to the survival of families, communities and humanity results in the process. Such minimisation and diminishing has often resulted in the subjugation of women's knowledges to facilitate male dominance. Evidence of such subjugation is exhibited in the ways that ongoing efforts at bushfire fighting remains gendered in the ways that it fails to incorporate women's concerns and knowledges. It is for this reason that a reclamation of the feminine principle in general is necessary and the experience of reproduction, specifically, is critical.

The uniqueness of the feminine principle in reproduction lies in the continuity that women maintain. Continuity here refers to the extended period of time and the sustained relationship that women maintain in the process. In the reproduction process, women are not only concerned about the conception and birthing of the young but also their nurturing to maturity. Right from conception, women maintain a special bond with the baby. This bond is internal and emotionally laden in ways that men have no way of experiencing. The challenges of labour and the arrival of the infant intensify this bond. As a result, on arrival women devote their entire life to nurturing the baby to maturity. In the process, special relations and connections are established between mothers and children in ways that are mutually reinforcing rather than destructive. This process has significant teachings for human relationship with nature.

There is no doubt that humans depend on nature. It is also indisputable that such dependence has resulted in the destruction of nature thus threatening human survival. The challenge then is how to depend on nature in ways that are mutually reinforcing rather destructive. This challenge requires that humans make concerted efforts at ensuring that their activities on nature are, first, re-generative. Deriving from the process of reproduction, re-generative activities include those activities that have at their centre the drive to multiply and increase nature's potential. For instance, in the search for firewood to cook the family meal, women under normal circumstances will harvest only dry wood which have completed their life cycle rather than wet wood that are still active. By such action, trees are allowed to grow to maturity. Indeed, if women were harvesting wet wood, they would only be able to harvest young trees, which their low level of technology will allow them. However, as victims of the destructive power of advanced technology, women have been compelled to participate in the process. Women have had to resort to fire to burn the bush so that they can harvest fuel wood more easily with simple cutlasses. Rather than harvest dry wood, women can be found actively cutting wet wood, which they dry for use as fuel wood. Women are also found cutting and burning wet wood into charcoal for sale or domestic use. Unless this cycle is broken the peril of not only nature but also humankind is what awaits all.

Second, human activities must be guided by the need to maintain nature for the continual support of human needs. Harvesting from nature without complementary efforts to nurture it back to life is anything but sustainable. The continual burning of the bush renders nature ineffective in sustaining itself. Trees that have been burnt over and over again through the annual wildfires die out, eventually rather than regenerate. When dead, they are only used up and are incapable of continuing their roles of supporting humans and nature. The food, medicines, and energy support that they have always provided for humans is terminated forever. The habitat that they have always provided for other life forms such as animals and plants is also destroyed for good. Hence, the destruction of the bush through burning becomes a violation of not just humans but also bio-diversity. Unless humans understand the need to maintain continuity, such violence will continue to be perpetuated. A return to the feminine principle teaches us to nurture rather than destroy. It also shows how such nurturing has mutual benefits and fosters the continuity of all life forms.

Conclusion

Discussions above have entailed assertions, interrogations and proposals that have been aimed are fostering sustainable wildfire management. Assertions have been made regarding the inefficacy of ongoing efforts at wildfire management and also on the stakes for women during the outbreak of wildfires. Ongoing efforts have also been critiqued for their gender biases in the ways that they exclude women's experiences. This has served the basis for the proposal for the inclusion of women's experiences and learning from feminine principles.

By learning from women's experiences, it has been suggested, humanity will be able to sustain life in the ways that women do. Humanity will need to return to the feminine principle that was lost in the quest for modernist development through scientific and technological advancement. The resulting death of the feminine, it argued, had resulted in the wanton destruction of the environment and the lack of appreciation of the relationships among all life forms. The return to the principle and to the lessons that it teaches in the form of care, interdependence and continuity is meant to not only rescue humanity but also all life forms from annihilation. The effects of the lack of care, appreciation of interdependence and maintenance of ontological continuity are reflected in the reckless burning of the bush and destruction of the environment. Above all, the failure of ongoing efforts to stamp out such recklessness requires that efforts be re/channelled toward the centring of the feminine principle, where teachings regarding sustainability abound.

NOTES

1. Since 1983 Ghana has embarked on World Bank/IMF sponsored development planning. The West African nation has gone through various phases starting from the Economic Recovery Programme through Structural Adjustment Programme and now the Highly Indebted Poor Countries Initiative.
2. Gender Training Workshop organised by Action Aid Ghana representative and heads of decentralised departments in the Upper East Region of Ghana.
3. As part of background research on this paper separate visits were made to the two offices where conversations were held with officials.
4. Conversations with personnel of the Upper East Offices of the Fire Department and National Disaster Management Organization strongly support this claim.
5. This was a response to the question why female staff is not seen on fire scenes.
6. Analysts such as Irene Dankelman, Joan Davidson, Vandana Shiva and Maria Mies write extensively about the many ways that the manifestations of feminine principle in women's activities at the grassroots and in rural communities and the principle promotes sustainable development. In the last chapter of Staying Alive, Vandana Shiva demonstrates how the feminine principle protects nature.
7. Derives from the work of Descartes, whose search for certainty results in the separation of objective reality from subjective experience. His central is often framed as: "I think therefore I am."

REFERENCES

Alvares, C. (1994). *Science, Development and Violence: The Revolt Against Modernity*. (Oxford India Paperbacks). Delhi, India: Oxford University Press.

Apusigah, A.A. (2006). "Promoting Wildfire Management in Northern Ghana: Learning from History." *European Journal of Social Sciences*. (in review).

Apusigah, A.A. (2004). "Promoting Sustainable Bushfire Management in Ghana: A Gender Perspective." Millar, D., Apusigah, A.A. and Berinyuu, A. (eds.), *The Chief, The Forestor and the Fireman: Proceedings of the Bushfire Workshop of February 2004*. Tamale, Ghana: UDS/Care International.

Boserup, E. (1970). *Women's Role in Economic Development*. London, UK: George Allen & Unwin (Pub.) Ltd.

Canadian Council for International Co-operation (August 1991). Two Halves Make a Whole: Balancing Gender Relations in Development. Ottawa, ON: MATCH International Centre.

Crusius, T.W. (1991). A teacher's introduction to philosophical hermeneutics. (NCTS Teacher's Introduction Series). Urbana, IL: NCTE.

Dankelman, I. and Davidson, Joan (1988). *Women and Environment in the Third World: Alliance for the future*. London, UK: Earthscan (Pub.) Ltd.

Escobar, A. (1995). *Encountering Development: The Making and Unmaking of the Third World*. (Princeton Studies in Culture/Power/History). Princeton, NJ: Princeton University Press.

Ghana Government (2003). The Ghana Poverty Reduction Strategy: An Agenda for Growth and Development. Accra, National Development Planning Commission.

Ghana Government (1990). *Control and Prevention of Bushfires Law, 1990: P.N. D.C. Law 229*. Accra; Ghana: Government of Ghana.

Ghana Statistical Service (2002). *2000 Population and Housing Census: Summary of Report*. Accra, Ghana Statistical Service.

Hesse, A. (2000). Ghana. In Correa, S. (compiler). *Weighing up to Cairo: Evidence from Women in the South*. (28-44). n.a: DAWN.

James, V.U. (1995). "Introduction: Sustaining Women's Efforts in Africa's Development." James, V.U. (ed.), *Women and Sustainable Development*. (1-14). Westport, CT: Praeger Publishers.

Mohammed, J.A. (1990). "Situational Analysis of Women Migrants in Tolon-Kumbungu Area of Northern Ghana." Research Report (upublished).

Mies, M. (1993). "The Myth of Catching-up Development." Mies, M. & Shiva, V. (eds.). *Ecofeminism*. (55-69). Halifax, NS: Fernwood Publishers.

Mies, M. (1986). *Patriarchy and the Accumulation on a World Scale*. London, UK: Zed Books.

Millar, D. (2004). "Foreword: Bushfire and Rural Livelihoods in Northern Ghana." Millar, D., Apusigah, A.A. and Berinyuu, A. (eds.), *The Chief, The Forestor and the Fireman: Proceedings of the Bushfire Workshop of February 2004*. (1-5). Tamale, Ghana: UDS/Care International.

Ministry of Lands, Forestry and Mines (MLFM) (2005, August). *Wildfire Management Policy*. Accra, Ghana: Government of Ghana.

Nikoi, G. (1998). *Gender and Development*. (1993 Kwame Nkrumah Memorial Lectures). Accra, Ghana: Buck Press.

Noddings, N. (1992). *The Challenge to Care in Schools*. New York; NY: Teachers College Press, Columbia University.

Snyder, M.C. and Tadesse, M. (1995). *African Women and Development: A History*. Johannesburg, South Africa: Buck Press.

Shiva, V. (1989). *Staying Alive: Women Ecology and Development*. New Delhi, India: Kali for Women.

Shiva, V. (1993). "The Impoverishment of the Environment: Women and Children Last." Mies, M. and Shiva, V. (eds.), *Ecofeminism*, (70-90). Halifax, NS: Fernwood Publishers.

Shiva, V. (1993). "Women's Indigenous Knowledge and Biodiversity." Mies, M. & Shiva, V. (eds.), *Ecofeminism*. (164-173). Halifax, NS: Fernwood Publishers.

Tater, R. (2004). "Critique of Bushfire Laws in Ghana." Millar, D.; Apusigah, A.A. and Berinyuu, A. (eds.), *The Chief, The Forestor and the Fireman: Proceedings of the Bushfire Workshop of February 2004* (90-99). Tamale, Ghana: UDS/Care International.

Tronto, J.C. (1998). "What can Feminists Learn About Morality from Caring?" Sterba, J.P. (ed.), *Ethics: The Big Questions*. Malden; MA: Blackwell Publishers.

Tsikata, D. (2000). *Lip-service and Peanuts: The State Machinery for Women in Africa*. (National Machinery Series). Accra, Ghana: Third World Network.

Wasai, Z. (2004). "Development of a National Action Programme to Combat Dissertification." Millar, D.; Apusigah, A.A. and Berinyuu, A. (eds.), *The Chief, The Forestor and the Fireman: Proceedings of the Bushfire Workshop of February 2004* (46-50). Tamale, Ghana: UDS/Care International.

Taiwan's Rapid Development and Environmental Problems

12

Govindasamy Agoramoorthy and *Minna J. Hsu*

Introduction

More than a century ago, naturalist Alfred Russel Wallace was impressed with the ecological beauty of Taiwan and he wrote, "Among recent continental islands there is probably none that surpasses in interest and instructiveness the Chinese island named by the Portuguese, Formosa" (Wallace, 1880). Taiwan certainly is a biological microcosm and one can travel from a mountain peak to coral reef within just a day. Taiwan harbours about 4000 species of vascular plants, 61 species of mammals, 400 species of birds, 92 species of reptiles, 30 species of amphibians, 140 species of freshwater fish and an estimated 50,000 insect species including 400 species of butterflies (Hsu and Agoramoorthy, 1997). Six national parks, 18 nature reserves and 24 protected areas have been designated to safeguard biodiversity—the protected area covers 440,290 ha, which is 12.2 per cent of the total land area. However, Taiwan's rapid political, economic and industrial growth has put heavy pressure on the natural environment, especially over the last three decades.

Political and Economic Development

In 1895, military defeat forced China to cede Taiwan to Japan. Taiwan reverted to Chinese control after World War II. Following the Communist victory on the mainland in 1949, two million nationalists fled to Taiwan and established a government using the 1946 constitution drawn up for all of China. Over the next five decades, the ruling authorities gradually democratized and incorporated the native population within the governing structure. In 2000, Taiwan underwent its first peaceful transfer of power from the nationalist Kuomingtan Party to the Democratic Progressive Party. Throughout this period, the island prospered and became one of East Asia's economic 'Tigers'. The dominant political issues continue to be the relationship between Taiwan and China—specifically the question of eventual unification—as well as domestic political and economic reform.

Taiwan's economy is robust and ranks the 17th largest economy in the world. Agriculture contributes only 2 per cent to GDP, down from 32 per cent in 1952. Only 1 per cent of the Taiwan's population lives below the poverty line. Because of its conservative financial approach and its entrepreneurial strengths, Taiwan suffered little compared with many of its neighbours from the Asian financial crisis in 1998. Exports to China—mainly parts and equipment for the assembly of goods for export to developed countries—drove Taiwan's economic recovery in 2002. Although the SARS epidemic, typhoons, corporate scandals, and a drop in consumer spending caused GDP growth to contract to 3.2 per cent in 2003, an increasingly strong export performance kept Taiwan's economy on track, and the government expects the economy to grow 4.1 per cent in 2004.

Threats to Wildlife Populations

The island's only bat, the Formosan flying fax, was previously abundant on Taiwan's Green Island, became extinct in the 1990's due to hunting, deforestation and habitat alteration (Hsu, 1997). The largest cat, the clouded leopard has not been sighted for two decades and is presumed extinct. The Formosan black bear has been seldom seen in the wild (Hsu and Agoramoorthy, 1997). The sika deer that once roamed the coastal plains became extinct in the wild in the 1960's due to intensive hunting and habitat destruction. The mangrove forest is considered to be one of the most highly endangered forests in Taiwan. Despite the existence of national parks and nature reserves to protect biodiversity, pressure on natural habitat intensifies due to poor law enforcement and weak policies.

To make things worst, large numbers of wildlife have been released into the wild by Buddhist and Taoist religious groups that have alarmed even animal welfare groups lately. People in Taiwan are estimated to spend about US$ 5,914,490 a year to set free 200 million wild animals for religious reasons that ranged from insects to monkeys (Agoramoorthy and Hsu, 2005). Taiwan's two major traditional religions, Taoism and Buddhism, stress the importance of good deeds during a person's life and the religions decree that returning animals to nature is one of the ways to garner good karma and will earn an individual's salvation in the afterlife. These religions form 93 per cent of the total population in Taiwan. Temples usually provide religious freeing services of animals to the faithful. Since there is a huge market for this trade, various animals such as birds, fish, snake, frog, turtle, and even insects and monkeys are being captured from the wild by hunters or purchased from local markets to be released around the island's rivers, mountains, forests, lakes, and reservoirs.

Although Taiwan enacted the wildlife conservation law in 1989 (Agoramoorthy and Hsu, 1999), the enforcement of law has been weak and there is no policy to end the practice of religious freeing of animals, which at present is unregulated. The unscrupulous freeing of animals will certainly lead to alien invasions of non-native species, which is catastrophic to the ecological food chain. We have recorded about 75 species of exotic birds that are wild at present in Taiwan, which shows that alien species invasion has gone beyond management (Agoramoorthy and Hsu, 2005). This will further lead to extinction of endemic species, hybridism and irreversible alteration of the genetic makeup of biological communities in Taiwan. Politicians in general are not eager to come up with policies that would antagonize religious groups since they play a major role in elections and political decisions in Taiwan. However, Taiwan must act quickly to counter the unmonitored release of wildlife that has been ignored for decades by drafting specific regulations to ban such practices.

Developments Leading to Environmental Pollution

Taiwan's Environmental Protection Administration (EPA) estimates that only about 43 per cent of hazardous industrial waste is properly treated. Much of the untreated waste ends up contaminating soil and water via illegal dumping. In 1997, the government admitted to produce 146,000 tons of hazardous industrial waste. Based on recent data from 1000 large industrial firms, the EPA estimates that Taiwan produces over 18 million tons of technological solid waste annually; 1.47 million tons are considered hazardous. Only 600,000 tons are treated while the rest ends up in rivers and landfills. About 50,000 tons of toxic solvents are produced annually, 35,000 tons alone by firms in Hsinchu high-tech science-based industrial park, which is Taiwan's equivalent of the Silicon Valley.

In July 2000, 100 tons of toxic solvents were dumped into the second longest river, the Kaoping, which left 3 million residents in and around Kaohsiung city without drinking water for five days. In Taiwan's largest environmental crime, prosecutors charged the waste handler of the Kaoping affair with illegally dumping tons of toxic solvents into river systems across the island, a classical example of corporate greed (Agoramoorthy and Hsu, 2000). Since Taiwan doesn't have enough secure waste landfills, it does not have the ability to handle all the toxic waste it produces and the grim reality has spurred the government to seek poorer, cash-starved countries to dispose of the toxic waste for a hefty fee. In a high profile case, Cambodia returned 2,700 tons of mercury-laced waste produced by Taiwan in 1999 after several deaths occurred near the disposal site (Agoramoorthy and Hsu, 2000). The ship that carried the toxic cargo returned to Taiwan and anchored for a week but later disappeared without any trace! Developed countries including Taiwan continue to dump toxic wastes in developing countries in Asia often in the name of recycling (Agoramoorthy and Hsu, 2002; Agoramoorthy, 2006).

In 1997, local residents complained about pungent smells and polluted water in the vicinity of Hsinchu industrial park in northern Taiwan. When a nun at the nearby bible college fainted because of the smell, residents became outraged. The pollution was traced to mixed wastewater from the industrial park that was dumped into rivers and streams. There was no way to identify which company was responsible or what caused the incident. In 2000, the Hsinchu Department of Health tested the blood and urine of 255 local residents and 56 per cent of the people were found to have abnormal blood tests, and 41 per cent had abnormal urine tests. Thus monitoring of residents' health is vital to understand the consequences of industrial pollution and the government must enforce stringent laws and policies to regulate companies involved in industrial development to minimize environmental population.

Taiwan's waste production has not reduced in recent years, and in fact, the daily amount of waste going to incinerators has increased from 15,300 tons in 2000 to 21,150 in 2003. Intensive environmental campaigns are needed to alert and educate people to reduce the waste production. Taiwan's underground pollution ranges from biological contaminants to organic solvents to heavy metals to pesticides, and many of these substances have persistent toxicity that slowly accumulate in human body and can lead to major health problems such as cancer and birth defects. According to the EPA, nearly 20 per cent of all underground sources of drinking water in Taiwan are polluted. A variety of volatile organic compounds including vinyl chloride, tetrachloroethylene, dichloroethane and trichloroethylene are still present in the groundwater near semi-conductor factories. These chemicals are known to increase the risk of cancer. Taiwan's Department of Health released a statement saying that at least one resident has been diagnosed with cancer every 11 minutes (*China Post*, 2000). We recently investigated heavy metal pollution in various organisms ranging from fungi to flora and earthworms to snakes and found the terrestrial ecosystem in Kenting National Park, a protected forest reserve contaminated with heavy metals (Hsu et al., 2006).

Green Movements

When martial law was lifted during the late 1980s, democracy was born and it provided Taiwan's civil society with new opportunities to influence policy-making. Environmental movements slowly started to emerge but their impact on environmental policy processes was limited because the environmentalists' interests came second to the corporate conglomerates. Prior to the consolidation of democracy, environmentalists and pro-democracy activists voiced their concerns together. But, after Taiwan embraced democracy, the movements fragmented, which is similar to Brazil where fragmentation of reform movements benefited its opponents, especially the rich corporate businesses that were able to influence environmental policy to their advantage (Weyland, 1996). As a result, Taiwan's environmental movements distanced themselves from their allies in politics as the new democratic governments started to pursue economic growth at the expense of environmental policy reform.

As a matter of fact, the newly started Green Party contested alone in the 1996 elections in 52 electoral districts with 13 candidates and won over 100,000 votes in the national assembly elections with an average of 2.97 per cent of the vote. It was below the 5 per cent bottom line necessary to become an official political party in Taiwan. Green Parties worldwide promote ecological sustainability and lobby to root out social injustice; they follow four basic principles such as ecological sustainability, grassroots democracy, social justice and world peace (Bomberg, 1998). Taiwan's Green Party follows these basic principles and it continues to lobby for the abandonment of all nuclear power plants. The green party also opposes the expansion of the polluting petrochemical, steel, and cements industries. Although the party is ambitious to propose eco-friendly development projects, it will achieve significant success only when it emerges as a strong political force in Taiwan.

During the time of political liberalization, the strongest opposition to the Kuomintang government that had absolute power before Taiwan's democratic reforms came from the environmental and democratization movements. For example, the Democratic Progressive Party was a strong political advocate of the environmental and anti-corruption movement in Taiwan (Lee and So, 1999). However, when the party came into power in 2000 with the leadership of Chen Shui-bian, it failed to undertake major environment-friendly projects and also did very little towards environmental protection. An example is scraping of the fourth nuclear power plant, which was one of the major campaign promises of the Democratic Progressive Party. It occupied only minority seats in the Legislative Yuan (Parliament), so the decision to halt the project was opposed by the Kuomintang Party and later escalated into a political turmoil, causing a possible recall of the President. Subsequently, in February 2001, Chen Shui-bian's government revived the project to save his job and his cabinet. Furthermore, his government has been engulfed in corruption scandals in recent years that dissatisfied many environmentalists and environmental movements (Copper, 2003; *China Daily*, 2006). According to a recent report, Taiwan's Ministry of Justice recovered about US$ 865 million embezzled by local politicians and policy-makers. It prosecuted 3,241 cases involving 7,959 civil servants and elected officials mainly for taking bribes (*Taipei Times*, 2006). Street protests involving thousands of people paralyzed Taipei city on 15 September 2006 and citizens from all walks of life demanded the resignation of President Chen Shui-bian citing series of high-profile corruption scandals (CNN, 2006).

Environmental Campaigns

The environmental activism in the 1990's has the characteristics of broader participation of grassroots involving local people to pressure politicians and we would like to cite two campaigns namely the anti-industrial complex movement in Chiku Lagoon and anti-dam movement in Meinung. The Chiku town is located in the southwest coast of Taiwan and it consists of tidal mudflats, mangrove forest, aquaculture ponds, coastal forest and the Chiku Lagoon, which is the largest remaining lagoon in the southwest coast of Taiwan. The natural habitat of Chiku provides feeding area for hundreds of migratory waterfowl including the critically endangered black-faced spoonbill. More than half of the world's recorded population of black-faced spoonbill stays in the lagoon each winter. The conflict began during the early 1990's when two of Taiwan's most powerful corporation jointly proposed an industrial complex in the area. Originally planned to fill the entire Chiku Lagoon, the industrial complex is one of the largest (2,000 ha) industrial complexes being proposed in Taiwan in recent years. Environmental impacts from the proposed industrial complex became the focus of debate and resistance from local fishing community and environmental conservation groups that forced the government to reconsider development plans (Kondolf and Hou, 1998).

Meinung is a tobacco-growing community located in the mountainous region of Southern Taiwan where the majority of local residents are of Hakka ethnic descent. The controversy on the proposed Meinung dam began in 1992. The cost of the proposed dam was estimated to be US$ 4 billion (Liang, 1993). Upon completion, the dam will supply water to the region's growing heavy industries including the proposed industrial complex at Chiku Lagoon. In fact, planning for the dam began in 1980s, but local people were not aware of it until 1992. The government's failure to disclose the project to the local community has angered the communities and their leaders. A small group of local people including teachers and scholars initiated a movement that expanded into a broad coalition of residents and organizations. Between 1993 and 1994, the movement organized series of protests in Taipei and persuaded the Economic and Budget Committee at the National Legislature to eliminate the dam project.

Can Taiwan Achieve Sustainable Development?

Taiwan remained as a totalitarian state during the 1970s and early 1980s and public voices were seldom heard and policy-making was top-down rather than interactive, and economic development was accomplished at the expense of environmental destruction (Gold, 1986). The sustainable development consciousness only emerged when Taiwan transformed to democracy during the mid-1980s. However, the term 'sustainable development' was still new even in the 1990s (Tang, 2000). Despite the fact that the environmental policy-making was given some importance in the government when the Environmental Protection Administration (EPA) was promoted to a cabinet-level agency in 1987, the threat of pollution continue to increase over the last two decades. Even the implementation of the air pollution prevention fee in 1996 made the EPA appear too comfortable with the industrial sector since the fee was levied from automobiles (which emitted 7-10% of the targeted pollutants) instead from polluting industries (Tang and Tang, 2000)!

Taiwan's rapid development in the past few decades has been recognized internationally as an 'economic miracle' (Haggard, 1990; Gold, 1986). As a consequence, Taiwan continues to pay a hefty price in the deterioration of natural resources and environment (Agoramoorthy and Hsu, 2000). Thus it's time for Taiwan's politicians and policy-makers to develop strategies to sustain economic growth and protect the environment for future generations. Understanding the growing public environmental consciousness and strengthening channels of interactions between the government and society are the two main issues that the government of Taiwan can no longer ignore. The government also needs to develop new energy-efficient technology and innovation. These policies could provide market incentives for the business industry as well as minimize environmental stress in Taiwan's delicate island ecosystem.

The lack of experienced environmental scientists in government offices to make sound management decisions is one of the major problems. According to government regulations, one has to pass an examination to get into government agencies and this causes disadvantage to scientists who do not wish to go through the examination procedures. The government's officials therefore rely on advice from university professors to deal problems related to environmental conservation and management policies. Because academics do not bear responsibility for the implementation of policies, there is little pressure for their advice to take account of political realities and constraints (Hsu and Agoramoorthy, 1997). Furthermore, the directors of the national parks, nature reserves, environmental agencies and pollution control boards are often politicians who lack hands-on experience in managing environmental and biodiversity resources, which create often difficulty to make wise environment-friendly policy decisions.

Can people fit on this small island with a biosphere? Experts admit that wealth and over-consumption cause environmental degradation (Donohoe, 2003). If rich people can learn how to live simply, the poor people can simply live. India's national father, Mahatma Gandhi once said, "There is enough on this earth to meet everyone's need, but not anybody's greed." At present, humans use about 70 per cent of the world's temperate and tropical ecosystems to produce 98 per cent of their food and all of their wood products. Only 5 per cent of the temperate and tropical land area is uninhabited. Asia and the Pacific region have 23 per cent of the world's land area but 58 per cent of its people (Agoramoorthy and Hsu, 2001a). Taiwan's human population density has grown from 7 million in 1950 to 23 million in 2003 with a density of 638 people per km^2. The population explosion combined with the uninterrupted developmental activities and lack of law enforcement enhances environmental disasters. Economically progressive islands such as Taiwan, Japan and Singapore continue to imports all their wood and wood products from overseas including mega-biodiversity countries in Southeast Asia (Agoramoorthy and Hsu, 2001b). Rapid development is known to put heavy pressure on the environment even in rich countries such as the United States of America where 924 species fauna and flora were reported to be endangered in 1997 distributed in 2,858 counties (Dobson et al., 1997). Although Europe has added 10 million hectares of protected areas since 1982 but 52 per cent of its fish, 45 per cent of its reptiles and 42 per cent of its mammals in addition to 86 per cent of the coastlines are under threat (Global Environmental Outlook, 1997).

Taiwan's industries (30.3% of GDP) and services sectors (67.9% of GDP) play a major role in the economic development. Undoubtedly, Taiwanese technological and economic growth has led to an increasingly polluted society and the extinction of species (Taiwan 2000 Study, 1989; Hsu and Agoramoorthy, 1997). The government of Taiwan spends a large sum for national defence followed by economic development and education, and environmental protection still remains a low priority since politicians seldom recognize the importance of protecting the environment (Agoramoorthy and Hsu, 2000). Therefore, we hope that the ongoing struggle against environmental destruction and pollution will contribute to a reorientation of Taiwanese environmentalism through burgeoning green movements to ultimately persuade politicians so that appropriate policies, sustainable development programmes and educational programmes can be enforced to defend the environment from further devastation.

People tend to view earth's environment as something esoteric, but must realize that human lives depend on our natural environment. Real conservation consciousness can come about only through the emergence of an informed, educated and healthy citizenry who are able to place environmental conservation into social, political and economic contexts at local, regional and international levels (Agoramoorthy, 1997). Balancing economic development and environmental conservation are daunting tasks for future leaders of Taiwan. Thus environmental protection and sustainable development should be the bottom-line for politicians, policy-makers and people from all walks of life.

REFERENCES

Agoramoorthy, G. (1997). "Conserving Wildlife, International Education and Communication Approaches—Book Review." *Journal of Mammalogy*. 61: 1442-43.

Agoramoorthy, G. (2006). "Computer 'Recycling' Builds Garbage Dump Overseas." *Nature*. 441: 25.

Agoramoorthy, G. and Hsu, M.J. (1999). "CITES Implementation through the Wildlife Conservation Law of Taiwan." *Journal of International Wildlife Law and Policy*. 2: 64-68.

Agoramoorthy, G. and Hsu, M.J. (2000). "Taiwan Pays the Price for Growth, in Toxic Pollution." *Nature*. 408: 905.

Agoramoorthy, G. and Hsu, M.J. (2001a). "India's Imminent Water Scarcity is a Concern for People and Environment." *Current Science*. 81: 1155-56.

Agoramoorthy, G. and Hsu, M.J. (2001b). "Conservation should be High Priority in Singapore." *Nature*. 410: 144.

Agoramoorthy, G. and Hsu, M.J. (2002). "Hazardous Waste Dumping in India by Developed Countries should be Discouraged." *Current Science*. 82: 1415-16.

Agoramoorthy, G. and Hsu, M.J. (2005). "Religious Freeing of Wildlife Promotes Alien Species Invasion." *BioScience*. 55: 5-6.

Bomberg, E. (1998). *Green Parties and Politics in the European Union*. Routledge, London.

China Daily (2006). "Chen Shui-bian Baffled by Insider Trade Scandal." 15 May 2006.

China Post (2000). "A Cancer Case Every 11 Minutes in Taiwan." 22 December.

CNN (2006). "Crowds Protest Taiwan President." www.cnn.com/2006/WORLD/asiapcf/09/15/taiwan.protests.ap/index.html.

Copper, J.F. (2003). "Taiwan: Democracy's Gone Awry." *Journal of Contemporary China*. 12: 145-62.

Dobson, A.P., Rodriguez, J.P., Roberts, W.M. and Wilcove, D.S. (1997). "Geographical Distribution of Endangered Species in the United States." *Science*. 275: 550-53.

Donohoe, M. (2003). "Causes and Health Consequences of Environmental Degradation and Social Injustice." *Social Science & Medicine*. 56: 573-87.

Global Environmental report (1997). Global State of the Environment Report. United Nations Environmental Program.

Gold, T. (1986). *State and Society in the Taiwan Miracle*. M.E. Sharpe, New York.

Haggard, S. (1990). *Pathways from the Periphery*. Cornell University Press, Ithaca.

Hsu, M.J. (1997). "Population Status and Conservation of Bats (Chiroptera) in Kenting National Park, Taiwan." *Oryx*. 31: 295-301.

Hsu, M.J. and Agoramoorthy, G. (1997). "Wildlife Conservation in Taiwan." *Conservation Biology*. 11, 834-36.

Hsu, M.J., Selvaraj, K. and Agoramoorthy, G. (2006). "Taiwan's Industrial Heavy Metal Pollution Threatens Terrestrial Biota." *Environmental Pollution*. 143: 327-34.

Kondolf, G.M. and Hou, J. (1998). "Review of the Environmental Impact Assessment for the Binnan Industrial Complex, Tsengwen Coastal Plain, Taiwan." Report No. CEDR-98-001, Center for Environmental Design Research, University of California, Berkeley.

Lee, Y.S. and So, A. (1999). *Asia's Environmental Movements: Comparative Perspectives*. M.E. Sharpe, Armonk, N.Y.

Liang, H. (1993). "Fighting to Save Taiwan's Yellow Butterfly Valley." *World Rivers Review*. 8: 3.

Taipei Times (2006). "Ministry release Corruption Data." 06 July.

Taiwan 2000 Study (1989). Taiwan 2000: Balancing economic growth and environmental protection. The Steering Committee Taiwan 200 Study, Taipei, Taiwan.

Tang, C.P. (2000). "Democratic Administration and Sustainable Development: Comparing Environmental Impact Assessments in Taiwan and Hong Kong." *Issues and Studies*. 16: 17-35.

Tang, C.P. and Tang, S.Y. (2000). "Democratizing Bureaucracy: The Political Economy of Environmental Assessment and Air Pollution Prevention Fees in Taiwan." *Comparative Politics*. 8: 81-99.

Wallace, A.R. (1880). *Island Life*. London, MacMillan & Co.

Weyland, K. (1996). *Democracy without Equity: Failures of Reform in Brazil*. Pittsburgh: University of Pittsburgh Press.

Waste Management Practices and Sustainable Development in Nigeria: A Case Study of Isolo Industrial Estate in Lagos State, Nigeria

13

Adewole Taiwo Adegboyega

General Introduction

The study area is located in Isolo/Oshodi local government area of Lagos State, Nigeria. This area comprises of food processing industries, Textile industries, Pharmaceutical industries, Insecticides industry and Paint manufacturing industries among others.

This study will be looking at both solid waste management and waste water treatment in the industrial estate and Lagos State in general.

The Federal Environmental Protection Act (1988) does not define "waste", however Waste as the term implies is any solid, liquid or gaseous substances or materials which being a scrap or being super flows, refuse or reject, is disposed off or required to be disposed as unwanted, this is Environmental law, the term assumes it's ordinary literal meaning unlike in the real property law, when "waste" is used as a term of art, having meaning completely different from its ordinary meaning.

One of the few statues in Nigeria, which attempts to define waste is the Lagos State Environmental Edicts 1985, there in Section 32, Waste is define as follows:

Waste includes:

(1) Waste of all description

(2) Any substance, which constitutes scrap materials or an effluent or other unwanted surplus substances arising from the application of any process.

The United Kingdom Environmental Protection Act 1990, re-enacting an earlier U.K statue, took this statutory definition a step further in section 75(2), it defines waste in these terms:

Waste includes:

(a) Any substance which constitutes a scrap material or an effluent or other unwanted surplus substance arising from the application of any process and

(b) Any substance or article, which requires to be disposed of as being broken, worn out, contaminated or otherwise spoiled.

One thing to notice is that none of the above definitions of waste give "value" to the elements considered there is no suggestion that the items, which constitute a waste, do not have value or is intrinsically useless. The word "unwanted" which appears in the definition although it introduces its own problem, does not necessarily, import a value element for a substance or material that may be unwanted notwithstanding that it has some value.

It has been observed that many industries in Nigeria do discharge their waste into surface water (oceans, seas and streams) more often than not without any form of remediation or treatment. This have several deleterious consequences such as using partially diluted, polluted water for irrigation, reduction in the food quality and quantity useful and beneficial aquatic flora and fauna. The resultant effects of this will lead to food insecurity, reduction in life expectancy at birth, increased hospital visitation by all age groups in the population.

Industrial pollution from over 5000 industrial facilities and perhaps another 10,000 small scale industries, some operating illegally within residential premises, is a growing problem in Nigeria. In places such as Kano, Kaduna, Port Harcourt, Warri and Lagos, coloured, hot and heavy metal-laden effluent, especially that from the textiles, tannery, petrochemical and paints industries are discharged directly into open drainages and channels constituting severe damage to water users down stream, also disturbing is the practise where sewage/septic tanks are in their backyards, threatening the water quality of innocent neighbours who rely on their dugout wells for drinking water.

This paper will be sub-divided into two sections: Solid waste management and effectiveness of waste water treatment.

Waste Management and Sustainable Development

The end of the 1980s saw a radical reappraisal of our concerns over resource availability and use, the environmental consequences of resource exploitation and the relationship between the environment, poverty and economic change.

This re-appraisal has given rise to a new approach to environment and development issues an approach which seeks to reconcile human needs and the capacity of the environment to cope with the consequences of economic system. This approach is called sustainable development.

Sustainable development is an implied development without destruction, it is the judicious use of non-renewable resources for the present and future generations, that is non-renewable resources must be used at a judicious rate, neither too fast nor too slow and to ensure that the natural wealth that they represent is converted into long-term wealth as they are used.

Scholars, in Nigeria succinctly put it that sustainable development without jeopardizing future development, meaning that in our efforts to explore and exploit the natural resources to serve us, there is an obvious paradox evident in the need to ensure economic development, while protecting the environment.

It is important to note that (for sustainable development) there must be a balance between levels of development and the stock of natural resources, that is, development must be at a level that can be sustained without prejudice to the natural environment or to future generations.

Therefore, what this paper is driving at is, if there is to be Sustainable Development in waste management in Lagos State, the availability of land (for landfill), human resources, plant and equipment and other tools including capital must be readily available. Now and in the nearest future to clean up our environment of all types of waste. Taking into consideration both physical and population development of the state.

On the other hand, management may be defined as the use of people and other resources for the achievement of organizational goals. Therefore, waste management could be define as meaning the collection, keeping, treatment and disposal of wastes in such away as to render it harmless to human and animal life, the ecology and the environment generally.

The Issues of Waste Collection and Disposal

Domestic Waste Management, Collection and disposal have always been a universal problem. According to studies, it was noted that for years, the major problem in Israel (especially in Ramat Hovar) was the accumulation of tens of thousands of tons of organic wastes. Also in the U.S. until the 1970's Federal Agencies had little authority to regulate hazardous and solid disposal often took in an unsafe manner at landfills or in inclined Lagoons, with some wastes simply dumped on the ground or in surface waters, as reported in Environmental Quality, 1991.

Refuse and domestic waste will not constitute a strange sight to Lagosians whose streets are littered with tons of garbage from animal to human carcasses.

At present, private sector waste disposal operators diligently visit homes and carry away refuse bags, load them into waiting trucks and cart them away for final disposal. "That is good" the residents say. They, however, worry that a lot of littering goes on in the environment and the streets and avenues may have been ignored and not cleared.

Study has revealed that household account for about half of the solid wastes generated i.e. by weight in the third world cities, which includes Lagos. It has also been noted that domestic waste disposal management has received considerable attention not only in Lagos State but Nigeria generally. Despite this laudable attention, collection, disposal processing, treatment, recycling and utilization have defied solution.

The major problem caused by wastes to the environment is pollution characterized by various types of solid wastes which include the paper, textile plastic, metals, glass, bone, wood, vegetal matter and food remnant of multiple consistency.

It has been pointed out that the generation of waste materials is a problem that is not peculiar to Lagos alone. This problem is prevalent the world over as noted earlier. This problem is not peculiar to the third world alone but cuts across the industrialized countries of the world where the pollutant effect of domestic and industrial wastes have caused considerable concern to environmental scientists. Our problems emanate from solid waste essentially. There are wastes from discarded materials generated from domestic and community activities or from industrial, commercial and agricultural operations.

Major Classes of Solid Wastes

Municipal solid wastes generally can be classified in terms of three major sources of generators: residential, commercial, and industrial. Sometimes, institutional sources are separated from commercial sources and, thus a fourth source is referred to as institutional. In the traditional scheme of classification, residential (domestic) solid waste consists of household garbage and rubbish, or refuse. The garbage fraction is mostly in the form of wastes derived from the preparation and consumption of food (e.g., meat and vegetable scraps). An alternate term commonly used to describe the garbage faction is "putrescibles". In the traditional scheme, all wastes not classified as "garbage" are classified as "rubbish". The major constituents of rubbish includes glass, metal, and plastic wastes, yard and garden debris, wastepaper and paper waste.

It is against this backdrop that this paper intends to review the impediments to effective and efficient waste management in Lagos State for sustainable development in Lagos State.

Percieved Causes of the Intracable Waste Problem

There are many perceived causes of the intractable waste problem in Lagos State among which are:

1. Waste Disposal habit of the people
2. Attitude to work
3. Lack of adequate equipment, plant and tools necessary for waste disposal and collection
4. Corruption

5. Overlap of function of the state enforcement and waste management agency
6. Population effect.

Waste Disposal Habit of the People

Ignorance coupled with poverty may be adduced to the habit of most people in Lagos State especially in the densely populated areas of the state. It beats one hollow to see a man defecating in broad daylight on the side of the Highway or a woman with her wrapper pulled up doing her thing on the sidewalk or gutter in full glare of the public or where a man or woman parks his or her car and throw waste on the street, then one begins to wonder the reasons for these dirty habits of our people.

Nigerians are permanently accustomed to dirt. This is the truth; Evidence of this can be seen everyday by way of indiscriminate discharge of garbage into drains and at times on the highways.

In a survey carried out on Lagos Waste Management Authority (LAWMA), which is rested with the responsibility of providing facilities for refuse collection; in all the streets within the state and to cart away the refuse bins daily does not have refuse collection bins in most of the streets in Lagos State. Hence the irresponsible dumps of waste on the streets. Also, the frequency of carting away the refuse was not regular where the public refuse bins are found.

In another survey carried out. It was succinctly put it that the volume of municipal waste piled up (for disposal) will be influenced by nearness to disposal sites, accessibility transportation facilities street layout, composition of wastes methods and individual attitude. Our individual attitude to waste disposal in Lagos State leaves more to be desired.

A situation whereby a landfill that has been closed to the public is still being used as a dumpsite calls for questioning. Also where waste is placed on the road median, gutter side, inside gutter and roadside does not augur well for effective waste management.

Despite the facts that illegal communal waste dumps indiscriminately located in public places have been officially cancelled yet several illegal refuse collection points, were indiscriminately created by residents which pose health hazard and loss of environment aesthetics.

Attitude to Work

Douglas Macgregor's Management Theory X that claims that man is inherently lazy and does not like to work unless he is coerced, forced, chastised, beaten and driven like a slave seem very applicable to the average.

Nigerian Workers Attitude to Work

In Nigeria for instance, employee productivity is low when compared with available human and natural resources, the low productivity is due to certain factors including sociological factor, which is felt in the manifested lack of a sense of belonging in an organization, and the tendency by employees to perceive a job as another's business.

This negative attribute to work has negatively impacted on the waste management efforts of the state government poor attitude to work, poor coordination and inadequate communication among workers and the institution saddled with solid waste management responsibilities due to bureaucratic impediment and administrative hitches have resulted in chaos, confusion and ineffectiveness in delivery of many urban public services.

Inadequate Vehicles, Plant & Equipment and Tools Necessary for Waste Management towards Attaining Sustainable Development

Waste disposed or deposited at designated points of collection has to be transported either to the transfer loading station where sorting is done or to the incinerator facility or sanitary landfill or the final disposal point. It was further noted that for effective and efficient collection system, there must be enough and well maintained equipment such as trucks, tippers, pay loaders, bulldozers, road sweepers, compactors and others. In Lagos State municipal solid waste transportation, collection and disposal has been epileptic, since the state realized its responsibility to the environment.

A lot of problems such as inadequate number of vehicles, lack of spare, dearth of fund, poor technical know-how, poor maintenance culture, insufficient funding and lack of motivation has bedevilled the agency responsible for the disposal and collection of waste *in* the State.

The heaps and stretch of refuse which adorn our roads pollute the Environment and disfigure the landscape are nothing but the result of inefficient waste collection and disposal management method.

Corruption

Corruption is a canker worm that has eaten deep into every fabric of the Nigerian society. This we may not deny except to our collective demise and peril. The collapse in most of Lagos State Waste Management Authority infrastructure in the state may be traced to this Menace (Corruption).

It has also been reported in some instances that market women have had to bribe the Lagos State Waste management agency operatives before waste could be removed from market place. Also the Private Support Participants, truck pusher and scavengers have been known to bribe official before they can be dispose their waste at designated points, this has led to illegal dumpsites springing up at different points of the state creating bottleneck to the already chaotic situation of waste management.

Overlap of Function of the Enforcement Agencies

Achieving Sustainable development is inextricably linked to the promulgation and establishment and enforcement, regulations, legislation and control criteria on environmental management and pollution control, but an overlap in the agencies responsible for effective enforcement of the various laws may create problems for effective management.

A situation where you have Lagos State Waste Management Authority that is saddled with the responsibilities of waste collection and disposal grappling with Local Government authorities, Kick Against Indiscipline, Lagos State Environmental

Protection Agency, the Police and other enforcement agencies in the state may not argue well for effective enforcement and sustainable waste management.

For us to have effective waste management and sustainable development in terms of waste collection and disposal effort in Lagos State, the enforcement mechanism should be left with only one organization where many agencies must be involved, their role must be clear-cut and well spelt out. However, there must be areas where Lagos State Waste Management, therefore, is also sometimes overlaps within one unit of government.

The enforcement of environmental laws in Nigeria generally has been problematic. Apart from the axiom that laws do not operate in a vacuum, the management and regulation of the environmental laws has been beset by a host of problems, and has met with very limited success. These problems that hinder the enforcement of sanction on violators of the environment are political, social and economic.

It is, therefore, clear that any effort towards a sustainable legal framework for successful enforcement, avoidance of overlap of environmental laws must come to terms with these issues, as a positive step towards the protection of the environment through effective waste management. The people who live in the environment and the entire process of sustainable development must be given due consideration. This overlap and effort made to plug the loophole. In a situation as presently obtained where the local government cannot cart away waste from their backyard, the other agencies should be able to give a helping hand and be their brother's keeper.

Population Effect on Waste Management

Population growth has always effected waste generation, collection and invariably disposal due to population growth and higher standards of living, Municipal waste has been increasing in the order of 5 per cent annually in Israel.

Human population growth might be seen to be the root of virtually the works of environmental problems. The reduction of the earth resources has been closely linked to the rise in human population.

The population of Lagos State rose from 1,443,569 in 1963 to 5,685,981 in 1991 and by estimation, the population hit 6,947,191 in December 1996. It has been hinted in some quarter that the present population of Lagos State has gotten to the 15 million marks. This has imparted negatively on both the environment and waste generation in the State.

As reported in UNCHS (Habitat) (1987) global report on human settlements, urban growth and development vary spatially, from country to country, region to region and continent, for instance in Africa in 1980, Southern Africa was the most urbanized (49.0%) as opposed to Eastern Africa which is the least urbanised with 15.1 per cent of total population.

Nigerian urbanization trend has also been spectacular, to say the least for instance in 1921 there was no city in Nigeria with a population of over 1,000,000, while only 10 towns had population of 20,000 and above each.

In 1931, 24 towns had urban population status (in Nigeria, town with a population of 20,000 and above are regarded as urban), with two having up to 100,000 and above population. By 1984 estimates showed that there were 356 towns with population of 35,000 and above. While 62 towns with population of 100,000; 14 with 500,000 and above population respectively.

In fact, by 1984, Lagos had attained a millionaire city status conclusively, between 1984 and year 2004 (a spate of 20 years) the population of Lagos has increased by 19 million people this explosion in population growth has negatively impacted on the management of solid waste in the State.

Lagos State, which is the most densely populated state in Nigeria due to its commercial activities, the quality of waste generated in the state is in proportion to population size—as population increases so also waste generated also increases. There is a significant relationship between population and waste generation in Lagos State.

With many other cities in the urban developing world, cities in Nigeria (especially Lagos) are faced with the twin problems of population increases and rapid expansion. These phenomena have no doubt brought increasing strain on urban infrastructural facilities. One area in which this strains has become obvious is in Waste Management where the existing system appears to be incapable of coping with the mountain load of waste generated and heaped on the surface.

In conclusion, population growth goes hand in hand with increased pollution and environmental decay.

Government Agencies Responsible for Waste Management

There are several government agencies in Lagos State responsible for the environment.

These among others are:

(1) Federal Environmental Protection Authority (now Ministry of Environment)

(2) Lagos State Environmental Protection Agency (LASEPA)

(3) Local Governments

(4) Lagos State Waste Management Authority

(5) Private Sector Participation (PSP), Non-government

These agencies in one form or the other are empowered to enforce or contribute to the enforcement of environmental laws in Lagos State directly or indirectly. Waste management in Lagos State, ordinarily falls under the portfolio of Lagos State Waste Management Authority, Local Government and the Private Sector Participation (PSP) initiative.

Focus on Lagos State Waste Management Authourity

The Lagos State Waste Management Authority (LAWMA) is the agency charged with the responsibility providing solid waste management services in the state. The organization was established in 1977 due to rapid growth of population in the state and lack of institutional and financial capabilities of the Local Government Council.

In 1997, when it became clear that government, alone cannot shoulder the responsibility of solid waste management as a social service, it introduced the involvement of the private sector through a pilot scheme named PSP (Private Sector

Participation) in domestic waste, collection in two local government areas:

> Although the scheme faced a myriad of complaints at commencement, it still formed the basis of introducing current PSP, which was re-introduced in December 1999. But around, the PSP was assigned the responsibility of transportation in all the twenty recognized local government in the State.

The main objectives of this scheme include:

1. Creation of employment opportunities.
2. Allow Local Government to take over the responsibility of PSP in their various localities and take charge of the collection, transportation and disposal of solid waste in their local governments to landfill sites.
3. Free Lagos State Waste Management Authority as it were to concentrate on the collection of Industrial/commercial wastes and management of its there landfill sites and transfer.

Despite the second objective, it is sad to note that the Local Governments have failed in their responsibility because Lagos State Waste Management Authority is still responsible for the collection of 80 per cent of domestic waste generated in the state.

Therefore, at present, LAWMA's role in solid waste management is as follows:

(1) Collection of Domestic Waste

(2) Collection of Industrial Waste

(3) Management of Landfill Sites

(4) Management of Transfer Loading Station

(5) Maintenance of Dumpsites/Landfills

At the inception of Lagos State Waste Management Authority as Lagos State Refuse Disposal Board in 1977, there were five existing landfill/dumpsites in operation within Lagos metropolis namely:

1. Pelewura (Adeniji Adele), Lagos Island Local Government now used as a transfer Loading station.

2. Gbagada, Kosofe Local Government Area.
3. Isolo, (Oke-Afa) Oshodi-Isolo Local Government Area.
4. Achakpo, Ajeromi-Ifelodun Local Government Area.
5. Ojota, (Odo Iya Alaro), Kosofe Local Government Area.

These sites were open swamps progressively reclaimed with refuse. The environmental implications, of waste management activity were of secondary concern and importance to the waste being disposed; the five sites have been closed to operations dating back to 1996 but are still in use illegally.

A World Bank assistance in 1988 led to the opening of the present, three legally operated sites at:

1. Olushosun in Ikeja Local Government Area
2. Abule-Egba, in Alimosho Local Government
3. Solous, also in Alimosho Local Government.

Legal Framework for Waste Management and Enforcement

A solid waste management strategy cannot be implemented without the support and guidance of a legislative framework. Legislation should be carefully drafted and should include a series of ordinances and regulation aimed at managing the solid wastes the legislation should not only include procedures for controlling emissions, but they should also deal with methodologies for monitoring and enforcing the regulations. Ideally, the legislation would be succinct, fair, and cost-effective. In addition, a thorough analysis of the potential impacts of the laws should be carried out prior to their issuance in terms of public health protection, environmental preservation, and overall influence on economic development.

The task of halting degradation and reversing pollution trends hinges on the enactment of Action Enforcement Policy backed by appropriate legislation that are rightly enforced.

A situation where several agencies lay claim to the enforcement of environmental laws in the state is not tenable and can only create confusion and chaos. There is need for capacity building Environmental managers, officers, enforcement agents, etc. must get rained in adequate numbers and quality. There is need to incorporate a "bottom up" and participatory approach to environmental management in Lagos State.

In other words the ordinary man on the street must be allowed to assume responsibility for managing his environment and participating in the neighbourhood, local and state efforts at managing waste disposal and collection.

The Public Health (Lagos) Act, 1988 Cap. 165 deals with environmental sanitation at the domestic level, while the criminal code Vol. V, 1990 (Laws of the Federation of Nigeria) contains penal provision for the detrimental vitiation of the atmosphere: There are the environmental sanitation edicts, which implement existing Public Health Laws and the Town Country Planning Laws.

All these laws reveal a common concern for and control of environmental sanitation issues, including: refuse and waste disposal, industrial and commercial pollution, enhancement aesthetic quality of the environment, etc. These are beautiful legal documents, which in the opinion of this paper should be vested in one Authority for proper implementation and enforcement.

To crown it all, we must endeavour to shun corruption or gratification in the discharge of our duties. We owe the environment this to build a virile, sustainable and manageable waste collection and disposal system for our collective survival.

The present form of orthodox environmental management lacks social equity and has so far proved to be too expensive for the government. Therefore, it must be integrated with popular environmental management to become participatory waste management. Participatory waste management will have the people as its focus, and uphold the primacy of their definition of needs perception, knowledge and values.

In this connection; it will be necessary to remove the obstacle to effective popular participation in waste management policies at the local level, such as illiteracy, ignorance, and certain, negative, cultural and religious beliefs and mass poverty. That prompts our people to defecate on the street or throw their waste carelessly. It will also be necessary to emphasize that except the basic needs of the people: food, shelter, work, etc. are provided, enforcement cannot succeed.

Thus the ordinary people who had been excluded from decision affecting the basis of their livelihood must become the subjects of policies and laws, within the framework of human-centred strategy of waste management in Lagos State (especially the market men and women).

The process of enforcement itself should be fair and just and not unduly skewed in favour of the rich and powerful element, in society. It is conceded that the infusion of equity consideration into waste management law must be generated from society itself to greatly enhance the process of sustainable enforcement.

It is, therefore, imperative for the process, of enforcement to contend with the issues of equity and must be socially responsive if it is to be sustainable within a social and democratic context.

The concept of sustainable development must be ingrained in the minds of policy-makers, lawmakers and the people themselves. It should be underscored that the notion sustainable development is central to political and social considerations, which in turn will then development objectives, for the use of future generations.

A situation where the Local Government and Private initiative is managing the waste process uncoordinatedly with the central body (LAWMA) does not allow for an efficient and waste management in Lagos State.

Way Forward for Waste Management in Lagos State Towards Sustaintable Development

Some of the major problems confronting and militating against an effective management and sustainable development of waste collection and disposal in Lagos State have been identified in this paper to include among others:

1. Population growth.
2. Waste Disposal Habit of the people.
3. Attitude to work (of those rested with the responsibility of collecting and disposing the waste).

4. Lack of adequate equipment and plant and other tools necessary.
5. Corruption.
6. Overlap of function enforcement agents.

The entire above highlighted problem cannot allow for effective management and sustainable development.

Therefore, it is against the background Issues raised in this paper that the writer will try to address the effective waste management system to be adopted for a sustainable development.

Mitigating Measures

On the issue of population growth, I will strongly suggest a rolling plan on waste management for Lagos State. A situation whereby the population of the State is projected for another 20 years and effective plans put in place to monitor and control waste expected to be generated during that period.

Expanding recycling programmes can help reduce solid waste pollution but the key to solving severe solid waste problems lies in reducing the amount of waste generated. It was noticed in the course of this investigation that only the landfill system of waste disposal is being generally adopted in Lagos State. Whereas in other places for example, there are several methods of waste disposal used to ameliorate and mitigate the issue of population effect on waste management.

Such systems that can or may be adopted for the state in the rolling plan are:

1. Recycling
2. Bio treatment
3. Incinerations
4. Neutralization
5. Secure Sanitary landfill
6. Composting

Also International cooperation should be sought to learn how other countries have effectively managed their waste collection; handling and disposal. There are several International Conventions on Waste Management; The Lagos State government should size the opportunity to apply for assistance in an effort to mitigate the looming disaster posed by population explosion in the state in terms of waste generation and disposal.

Remediation through education is also necessary. People of the state should be educated on the need to reduce the amount of waste generated. The Lagos Government should fund LAWMA to provide adequate collection bins in most areas of the metropolis and hinterland to forestall the wrong habit of throwing waste anywhere and everywhere, creating illegal dumps and doting and adorning the major streets of the state with wastes.

A sustainable development strategy action plan using a consensus-building approach should be formulated between government and other stakeholders to conform to the rolling plan. These should include national government the private sector, academics, environmental planners and experts, and non-governmental organizations to produce, an action plan document that may last 20 years before review. This document shall be published and made available for public consumption and fully implemented by all concerned.

As earlier mentioned, the waste disposal habit of the people may change if government stopped paying lip service to the serious issue of waste management. The availability and nearness of disposal sites will greatly enhance and improve the habit of dumping waste "anywhere and everywhere". Research and Development into areas of better waste handling method may also go a long way to assist in elaborating and interjecting the situation.

Nigeria has very little or nothing to showcase for as her achievements in the area of proper waste management. The military era saw the unsightly heaps of refuse litter the streets and drains of virtually all the hamlets, villages and cities from Lagos to Onitsha, and all other parts of Nigeria. Heaps of garbage is also commonplace along major roads, riverbanks, and ravines and in excavated areas, particularly places excavated to obtain sand for road construction.

In States where there is organized refuse collection, such as Lagos; the disposal of such wastes is usually an open dumpsites, located not far from living areas. Such dumpsites (called landfills) are not provided with environmental safeguards, and the leachates from them percolate freely into streams and the groundwater system.

With the exception of a few places, Nigeria cannot pride herself over having a functional sewage system. Industrial effluents of all types (both toxic and non-toxic) are discharged freely; into, surface and groundwater sources.

Waste is allowed to pile up before it is ordered to be cleared with military dispatch and automatic alacrity.

This leaves room for corruption and does not allow for effective waste management but that was the situation during the military era.

In developed countries, industries are compulsorily made to discharge all it effluents into only license on-land disposal sites, where such effluents are treated prior to re-use, recycling or discharge into streams or other approved places.

There are no such controls in Nigeria and where they exist they are not enforced and most industries discharge raw, untreated and highly toxic liquid effluents into open gutters, drains, streams, lakes, estuaries and lagoons. A visit to Osolo way in the Oshodi/Isolo local government area of Lagos State will attest to the argument put forward. The effluent from the Aswani Textile Manufacturing Factory and other industries around the area, are only discharged on the major road leading through the area.

This contributes to the continuous wearing of that section of the road year round. Rendering the roa ; sometime impassable especially on the path of the effluent. Despite this, market women and men troop to this place every Tuesday of the week to buy and sell not mindful of the great danger locking around.

To most Nigerians wastes is simply a nuisance, full stop. They hardly give serious thought to tile polluting effects of wastes or their deleterious effects on human health. Increase in urban population and 'blind', haphazard industrialization has contributed a great deal to the generation of wastes in Nigeria in the Municipal areas of Lagos State, more solid wastes are produced than the generators can effectively cope with or manage.

The situation of unmanageable wastes in the cities appears to worsen with perceived increase in the income of the inhabitants. The slums and the shanty neighbourhoods, as expected, receive little or no waste disposal services from anywhere.

Urgent Solution for Waste Disposal Sites in Lagos State

This study will generally concentrate on two types of waste disposal sites: Landfills and Open dumps. *A landfill* is differentiated from an open dump in that the landfill is an engineered design, consisting of a variety of systems for controlling the impact of land disposal on human health and safety and on the environment. An *Open dump* is an uncontrolled system and has not been the subject of an engineering design. This is the type of waste disposal been adopted in Lagos State in generally but commonly refers to as "landfill".

For the purpose of this study, a waste disposal site is generally defined to consist of that portion of the site wherein wastes are buried, as well as any surrounding property within the boundary of the site, the surrounding property may serve as a buffer, support landfill-related operation and facilities (e.g., maintenance) or unrelated activities (e.g., recycling depots), or contain access routes and roadways.

Requirement for Setting Up of Landfill in Lagos State

Acceptable definitions of a modern landfill are based on the concept of isolating the landfilled wastes from the environment until the wastes are stabilized and rendered innocuous as much as possible through the biological, chemical, and physical process of nature. The main differences among definitions of a landfill involve the degree of isolation and the means of accomplishing it. Isolation includes prevention of water from entering the landfill, as well as isolation of discharges directly from the fill to the environment. The rate and extent of decomposition of the wastes are subcategory of the design concept of isolation. As opposed to the rate and extend of decomposition that occur in a fill that is designed to minimize intrusion of water, certain designs and operating procedures can be used to accelerate the decomposition of the wastes, e.g., recirculation of landfill leachates. The benefits of these designs (termed bioreactor or wet cell designs) include less time to reach biological and chemical stabilization and leachates of lesser strength. Detracting features of these designs include the added complexity of the design and operations compared to landfill designs lacking systems to accelerate decomposition.

Three basic types of practices and requirements for a landfill:

1. Consolidation of wastes into the working face; compaction of waste to conserve land resources; design and operation of the fill to control settlement, to optimize the chemical and biological processes (e.g., for landfill gas recovering), or both;
2. Covering the waste with cover material on daily basis to control the risk of hazards from exposed wastes; and
3. Control or prevention of adverse environmental impacts wastes disposed on land to soil, water, and air resources and of their subsequent impact on public health and safety.

A landfill must meet the above three key conditions regardless of stage of economic development of the country in which the landfill is located. However, meeting the three conditions may be technologically and economically difficult or impractical in Lagos State. Therefore, the short-term, or immediate, goal should be to meet the conditions to the extent possible under existing circumstances. The long-term goal should be to eventually meet all three of the conditions. This approach is recommended since the benefits associated with a modern sanitary landfill are realized only to the extent that a land disposal facility fully meets the three basic conditions. The most important condition is the prevention of negative impacts on the public health and the environment.

In conclusion, knowledge of the quantities and characteristics of the wastes to be landfilled is fundamental to the proper design and operation of a landfill. Among other things, these parameters influence or control many aspects of the landfill system over its lifetime, including the annual rate of filling, the required volumetric capacity of the fill, production and characteristic of gas and leachates, and environmental impacts.

The Role of Scavengers: The role of scavengers is very important in the planning, implementation, and operation of land disposal sites in Lagos State. First, the occurrence of scavenging between the point of waste generation and the disposal location influences the quantities of waste that will be disposed, therefore this aspect of scavenging must be taken into account during the process of estimating waste quantities and characteristics in Lagos State.

Secondly, scavenging is a widespread occurrence at existing land disposal in Lagos State though this is opposed in most developed countries and is to be expected at new disposal sites unless policies and/or programmes are implemented to prevent the practice.

Scavenger are normally part of the socio-economic structure, their displacement from a disposal site can have many direct and indirect consequences. While unsupervised and uncontrolled scavenging is detrimental to the health and safety of the scavenger, as well as personnel operating the facility, the exclusion of scavengers from disposal sites is not necessary if their activities are managed and controlled.

Scavengers help in the recovering of valuable resources that would otherwise be disposed as waste, the reduction of problematic materials entering a landfill and the subsequent adverse effect that they have upon landfill operation and performance, for instance whole tires, toxic materials, etc. and the reduction of the quantities of materials destined for land disposal in order to conserve resources, e.g., land, and water and air quality.

Reason why Lagos State Adopt Open Dumps: In the case of developed nations, the degree of isolation considered necessary to protect the environment and human health and safety usually is much more than would be technically and financially practical in many developing countries including Lagos State. In the case of many developed countries, the high level of containment requires a complex and expensive engineered system. Therefore, the main reason why Open Dumps is adopted in Lagos State is due to the technical, financial, complex and expensive engineering system involved in setting up a Sanitary Landfill.

Current Trends in Urban Waste Management in Lagos State: Recently the Lagos State Government announced that they plan to return the management of urban wastes to the Local Government Councils; this is unacceptable to most people because the Local Councils have failed woefully in the past in this aspect.

In developed economies, Council authorities are rich and large administrative areas. The Tokyo, New York and London Municipal Council Authorities are very rich. They have the capacity to employ the required technical manpower, with the necessary equipment to manage municipal waste. Beside the people living in these places has imbibed the right orientation and don't dispose of waste indiscriminately. It is crazy for anyone to compare the London Municipal Council with the Oshodi/Isolo Local Government.

As it will be known to all, Oshodi/Isolo Local Government as at today cannot handle or monitor air/water pollution! This is because the Local Councils rely wholly on Private Sector Participation or Contractors to evacuate wastes. These Private Sector or contractors only do the job when they are paid because the ability of the Local Council to pay them depends on the Council income profile. The income they generate internally is not enough to do all their assigned duties. In Lagos State, the revenue meant for the Local Council from the Federation Account is irregular.

In the State, where the Federal Government has withheld the Council funds for more than two years, what becomes of urban waste if the councils were to do it?

A typical city like Lagos generates tons of refuse on daily basis. Lagos has an estimated population of over 12 million people. We should assume on the average that everyone in Lagos State generates just 1 kg of refuse per day, which means in a day we have 12 million kg (10,800 tons). In a week, we have 75,600 tons; and a month we have 302,400 tons of refuse. To evacuate this at once using the standard 7-ton tipper, then each will do 4,320 rounds. All this is just for one month, multiply this by 12 and we have 3,628,800 tons of refuse.

The major effects of waste management on the quality of life:

Environmental Effects: The major environmental effects includes Air pollution, which includes odour, smoke, noise, dust, etc. Waste pollution—pollution from disposal site via flooding because of blocked drains and land degradation.

Health Effects: This includes: flies which carry germs on their bodies and legs and also excrete them, mosquitoes; breed in stagnant water in blocked drains in favourable location in cans, tyres etc. that collects rain water.

Rats: They spread typlius, salmonella, leptospirosis and other diseases they cause injuries by biting and spoil millions of tons of food. The refuse workers also faces some hazards which includes: parasite infection and infected cuts resulting from skin contact with refuse, other includes hazards on disposal sites; are injuries from glass, razor blades, syringes, tissue damage or infection through respiration, ingestion or skin contact.

Urgent Tips to Attain Best Practices

- In term of population, there is an urgent need for rolling plans and education in order to monitor and control waste expected in the state. There should also be a room for International co-operation towards achieving the rolling plans.
- Expanding recycling programmes through the activities of scavengers among others (Waste-to-wealth).
- Landfills management and control; Waste-to-energy programmes can be generated through the landfills (Generation of Methane Gas).
- There is need for purchase of updated equipment; there is an urgent need for well trained staff, vehicles, trucks, tippers, pay loaders, bulldozer and road sweeper, which must be backed up with well stocked maintenance store provided for spare parts for all equipment.

- There is a need for an organised refuse collection both from residential and industrial estates. There must be a disposal site in each street, closes and avenue nearest to the sources of waste, which must be accessible by everyone and the collection should be daily and timed in all nooks and crannies of the federation.
- The support of private sector and NGO's is also required most especially in the area of organising maintenance workshops and enlightened programme, which should include grassroots participation and input.
- The Federal and State Government should properly fund the various agency responsible for effective and safe waste management and make it the pivot of the agency in the various states to be responsible for collection and disposal of all types of wastes.
- There is a need for at least one domestic waste incineration plants in each Local Government Council constructed with a daily waste treatment capacity of at least 3000 tons, which can generate electricity through incineration so as to recover the energy out of the waste.
- There is need for more effective and up-to-date domestic waste transfer stations with at least a daily transfer of 1000 tons.

Environmental Sanitation Day in Nigeria

As discussed, the military era in Nigeria and the oil-boom brought about a perceived increase in the income of Nigerians, especially the government workers and city dwellers. Nigerians, before the dawn of urbanization and unprecedented increase in population were basically environment-conscious. The various communities ensured at all times that their streets, market-squares and ceremonial grounds were regularly leaned up. In the South, the assignment was carried out by the appropriate traditional age-grade everywhere was green and lush; the air everywhere was refreshing. However, with the passage of time our environment deteriorated principally through human activities thanks to oil-exploration.

The Federal Government in 1985 introduced the environmental clean-up campaign in a spirited effort to stem the rising volume of municipal and rural wastes, and to inculcate environmental consciousness and cleanliness in all persons resident in Nigeria. The campaign stipulated that all residents in the country should compulsorily clean up their premises, streets and gutters every last Saturday of the month some state governments even went further to introduce twice-a-month clean-up exercise.

The initial impact of the environmental clean-up was felt mostly in the state capitals and some local government headquarters. Before long, the tempo of the laudable campaign initiative dwindled, and the clean-up exercise failed to yield the anticipated dividend.

Garbage generated from the clean-up exercises was heaped along major roads and streets or at designated dumpsites inside the state capitals and local government headquarters. Often a lot of the garbage was left uncollected by the appropriate local authorities for hours, days, weeks or even months. The garbage naturally, decomposed-spreading diseases and also polluting the air with an unbearable stench. Incessant public outcry and media condemnation of the state of affairs often caused the waste materials to be re-located from the streets to somewhere on the outskirts of towns in river channels abandoned excavation pits, ravines, farmlands and open spaces. The re-location of waste or waste transfer typifies the military era as earlier explained. The refuse was merely transferred from one location where people were inconvenienced to another where people hardly bothered about its presence. What would one expect from a country where there are no adequate facilities for incineration, recycling, composting and landfilling?

The civilian administration of President Olusegun Obasanjo cancelled the monthly environmental sanitation even though some states are still carrying it out, having discovered that the exercise was not a successful venture. The failure of the sanitation day exercise campaign shows that the government did not do enough feasibility study before commencing the exercise nationwide and the issue of effective waste management goes beyond military orders.

The efforts made so far in Nigeria to clean up the environment have been sporadic and unplanned. This mode of operation, however is not peculiar to Nigeria alone, as International Financial Agencies have recognized the difficulty in designing appropriate strategies for a clean environment, as a problem of the developing countries of the world. This explains why these countries voraciously demand or search for the sophisticated technology developed by the white-man as a remedial measure. This may not ameliorate the problem, as spare parts for the replacement of broken components are hard to come by. This is a major problem plaguing; the LAWMA today. A well thought out plan that takes the culture, habits, attitude, population and other factors into consideration is required if the issue is to be frontally addressed for sustainable development of the country.

The writer *is* strongly imploring the Lagos State Government to take a drastic step as seen in the picture painted above, to arrest the hopeless situation of waste management in the state in general and the metropolis in particular.

Industrial Wastewater Management in Lagos State

Most of the properties of water and wastewater depend on the charged particles. If the suspended particles in wastewater samples all have positive charges, the flocculation of these particles will be difficult. Water molecules behave like tiny magnets, in that they are polar molecules. An effluent solution contains many kinds of ions, the concentration of these ion increases, as the solvent evaporates from the solution. The concentration of ion increases as more minerals dissolved in water. Any further increases in the ion will cause some of this solid to precipitate. The solid that precipitates first is the one that saturated at lower concentration.

The impurities in water from a well are different from the impurities in a stream. Water mountain stream is likely to contain more dissolved air than the well water and water that percolates through the soil is likely to contain more dissolved mineral impurities than the rain water that run into a stream after a heavy storms. Swiftly running stream carries more solid materials in suspension that slow moving or still water. Oxygen, carbon dioxide and nitrogen are the main gases that river and stream pick up from the air. Well water and stream both contain minerals. The amount and the type of minerals depend upon the solubility limit of the minerals and their rate of dissolving.

Water Pollutant

Water pollutant can be defined as any physical, chemical or biological changes in water quality that affects living organism adversely or make the water unsuitable for the desired use.

There are two sources of potential pollutants:

Point Sources: They are identifiable sources that include domestic and industrial discharges, point sources discharge pollutants from specific location such as drain pipes, dishes and waterfall. Point sources are discrete, identifiable and are relatively easy to monitor and regulate.

Non-Point Sources: They are usually scattered of diffused having no specific location where they discharge into a particular body of water. Diffuse source may result from a large number of individual point sources e.g., automobile exhausts, agriculture run-off, which includes soil erosion from the earth surface as a waste disposal site, which include municipal and industrial solid waste disposal facilities.

Types of Water Pollutants

Water pollutants can be categorised into seven major types that are highly varied and interrelated.

1. Disease Causing Organism
2. Oxygen Demanding Waste
3. Sediments or Suspended Materials
4. Organics
5. Inorganic Plant Nutrients (Nitrate & Phosphates)
6. Radioactive Substance
7. Heat (Thermal Pollution)

Disease Causing Organism: Pathogenic organism are the most serious water pollutant in term of human health worldwide and they account for a wide range of water borne diseases such as dysentery, typhoid, cholera enteritis, polio, schistosomiasis and many others the insect that has aquatic larvae transmits even malaria and fever. Altogether about 25 million deaths are blamed annually on this water related disease. The main source of this pathogen is from untreated or improperly treated domestic and industrial waste.

Oxygen Demanding Waste: The amount of dissolved oxygen in water is a good indicator of water quality. It also represent a kind of life water can support, e.g., a water body with dissolved oxygen of 8 ppm and above will support fish and other form of aquatic life, while water with dissolved oxygen lower than 2 ppm will support worms, bacteria, fungi and other types of decomposers. Oxygen is added to water by diffusion from the air especially when turbulent and mixing rate are high as well as photosynthesis of green plants and algae in water. Oxygen can be removed from water by respiration and chemical processes that consumed oxygen.

The addition of biodegradable organic waste such as domestic sewage, food processing, pulps paper waste, etc. to water, stimulates the consumption of oxygen by decomposer. The decomposition of this oxygen waste, by oxygen consuming bacteria deplete the dissolved oxygen in water leading to the death of the aquatic plants and with associated foul odour. The impacts of this organic waste on water quality are expressed in term of biological oxygen demand (BOD). BOD represents a standard test of the amount of dissolved oxygen utilized by aquatic micro-organisms over a period of five days.

Sediments or Suspended Materials: Sediments and suspended materials are the largest volume of water pollutant in most part of the world. Naturally, water bodies are always known to carry sediments to the ocean but the erosion rate in many areas has been greatly accelerated by human activities, leading to a corresponding increase in the sediment being transported. An estimated amount of 24 billion metric tonnes of soil sediments and suspended solid are contributed annually by erosion, run off from agricultural and grazed land toward surface water.

A major consequences of high sediments and suspended load in water is the obstruction of shipping channels, clogging of hydroelectric turbines and increase in the cost of purification of water. Water with high level of suspended sediments is less suitable for fish and other form of aquatic life. This prevents the penetration of sunlight and as a result plants cannot carry out photosynthesis and there is a resultant decline in the dissolved oxygen level. From aesthetic point of view, milky and cloudy water is less attractive for swimming, fishing and other recreational activities.

Organic Chemicals: This comprises soluble and insoluble gasoline, plastics, pesticides and solvents etc. several natural and synthetic organic chemicals are used in the chemical industries to manufacture pesticides, plastics, pharmaceuticals, and pigments. Many of these chemicals are highly toxic; exposure to very low concentration can cause birth defects, genetic disorders and cancer. They are resistant to degradation and toxic to organisms that ingest them. Contamination of surface water and underground water by these organisms is a threat to human health. The most important sources of toxic organic chemical in water are improper disposal of industrial and household wastes as well as run off from agricultural field.

Inorganics Plant Nutrients (Nitrates & Phosphates): Water-soluble nitrates and phosphates are known to stimulate plants growth. They are introduced into water through natural run off from agricultural land, industrial waste discharges, inadequate water treatment, food processing industries and phosphates in detergent. Water clarity or transparency is a useful measure of water quality. The leaching of toxic metals from the soil and rocks into water makes it unfit for drinking and irrigation.

Radioactive Substances: The radioactive substances may be derived from natural sources such as rock, soil uranium mining and processing, nuclear power generation and nuclear power testing. The major effect of exposure to radioactive substances is cancer and genetic defect.

Heat (Thermal Pollution): Raising or lowering water temperature from normal level can adversely affect water quality and aquatic life. Usually water temperature is more stable than air temperature and consequently aquatic life are adapted to rapid temperature changes, lowering the temperature by one degree can cause a problem to some aquatic organism and raising water temperature can have similar devastative effects on aquatic organism. The solubility of oxygen in water decreases when temperature increases. Human beings cause chemical pollution by alteration of the vegetation and run-off pattern as well as discharging heated water from industrial cooling system directly into the river and lakes. Fish and other aquatic organism that are adapted to a particular temperature range, can be killed by thermal shock. Rivers and lakes that have clear water but low biological productivity are said to be eutrophic meaning truly nourished.

Water Quality Standard Technology-based Effluent Standard

Standard for water quality protection may be divided into:

1. Effluents, which are technology based.
2. Water quality standard, which refers to water receiving the effluent (e.g. stream, rivers, lakes, estuaries, etc.) or being affected by the effluent. This is also called stream standard water quality or stream standard which reflects the assimilative capacity of the water body receiving the effluent and designed to preserve the recipient water at a certain minimum quality required for its designated use. Normally each water bodies are assigned to a class, e.g. class AA (Extraordinary), class A (Excellent), class B (Good), and class C (Fair).

Limits for discharge are set to maintain the quality of each water bodies within that class or to improve its quality to a higher class.

Stream standard is usually considered a policy goal or objective and it is usually non-enforceable. Technology based effluent standards are set for point source discharger. They resist the quality of pollutants in the effluent or set the desired degree of treatment, which is technologically and economically feasible.

It is noteworthy that the effluent standard is practically enforceable. The conventional pollutants control in effluent are pH, colour, turbidity, biochemical oxygen demand, chemical oxygen demand, oil and grease, suspended solid and toxic organic, which include pesticides, PCBs, dioxin and asbestos. The USEPA has identified 127 priority pollutants that must be controlled in effluents.

Generally, the identifiable source of wastewater from factory includes:

- Raw materials handling
- Manufacturing processes
- Defective product
- Expired raw materials and products
- Industrial energy utilization
- Bye-products industrial processes
- Non-processes waste: office, restaurant, cloak room, etc.
- Combustion processes
- Accidents

Factors that determines the generation rate and characteristics of effluent:

- Industrial category
- Manufacturing processes
- Plant capacity
- Product type

From the ecological perspective, factors that determine the pollution potentials of waste include:

- Waste generation rate
- Waste qualities
- Toxicity of the waste
- Ecological characteristics of the surrounding
- Presence or absence of pollution abatement

Industrialization remains an essential pre-requisite to bring development into any nation and both industrialization and development are intrinsically linked. But for development to be meaningful it must be sustainable, which means it must be able to meet the needs of the present population without jeopardizing the ability of the future generation to meet their own needs.

Environmental issues in industries are diverse, depending on the nature of raw materials, production technology and packaging. It must be emphasized that industrial process leading to the production of goods and services required the handling and manipulation of chemicals. However, toxic and their bye-product if not properly managed could result in deleterious effect on public health and harmful consequences for environments.

The Federal Ministry of Environment is charged with the full legal responsibility to control and overseas the state of Nigeria environments. Industrialization has afforded man a standard of living, solved many technological problems and in the process, enhances quality of life to an extent, previously imaginable. However environmental pollution, which has caused depletion of resources and impairment of health and other living resources, is conversely the adverse consequences of industrialization.

Result of Sample that was Randomly Collected in Isolo Industrial Estate

Five samples were randomly collected from each site. These were labelled for site A1 to A3 for site A, B1 to B3 for site B and C1 to C3 for site C. Samples were taken in three strategic location within the area; these were drawn in such a way to avoid aeration, contamination and to maintain the true proportion of suspended solid in the sample.

Analysis of the Samples

The physical, chemical, biochemical and biological analysis of the samples were analysed using both the chemical and instrumental methods of analysis.

The following were determined:

Physical Analysis

1. Temperature
2. pH determination
3. Conductivity Measurement
4. Determination of suspended solid

Chemical Analysis

1. Determination of acidity
2. Determination of alkalinity
3. Determination of chlorides of the sample
4. Determination of total hardness of the samples
5. Determination of Lead
6. Analysis of Zinc in the sample
7. Determination of Biochemical Oxygen Dissolved (BOD)

Microbial Analysis of the Sample

The analysis of the sample in Zones A, B and C for physical parameters such as temperature, colour, odour, PH, flow rate, conductivity, settable solids, suspended solid were carried out under the specified environment. All the above parameters were analyzed in triplicate and some statistical data were obtained.

The average pH of the sample are: A=6.67 and B=6.27 and C=11.84. Samples in zones A and B complied with Federal Environmental Protection Agency Standard of 6-9 but sample C which was 11.843 (Textile area) was outside the standard. This is in conformity with the research carried out by Federal Environmental Protection Agency on the effluent quality of some textile mills in Nigeria 1991.

The implication of this is that the streams and rivers at the receiving end will be grossly polluted. Sample A and B are off white in colour, sample C shows variation in colour i.e. from dark brown to yellow compared. This same sample also had pungent smell, which was an implication that the effluent was contaminated by ammonia and this could be very dangerous to human health and the aquatic life.

The conductivity of the entire sample was high (A=0.203, B=0.726 and C=5.73) $uScm^{-1}$; it shows that the effluent contained high dissolved inorganic salt, although there was no specification given by Federal Environmental Protection Agency. Other physical parameter, e.g., suspended solid (A=29.60, B=28.00 and C=37.66)°C, all complies with Federal Environmental Protection Agency Specification for effluent e.g. highest suspended solid was 0.523 mg/l for site A and Federal Environmental Protection Agency was 30 mg/l, highest temperature was 37.66 °C for site C but Federal Environmental Protection Agency was 40 °C. Dissolved solid highest was 262.08 mgl^{-1} for C and Federal Environmental Protection Agency was 2000 mgl^{-1} maximum. Settleable soil (site A=0.229, site B=0.223 and site C= 0.4123) mgl^{-1}.

The alkalinity, acidity, chlorides, total hardness, magnesium hardness, calcium hardness, chemical oxygen demand (COD), zinc and lead, were analyzed in triplicate, the average was computed.

The alkalinity of sample A and B show Zero levels as per the chemical methods used in this project, but sample C was 135.17mg/l, and these are within the Federal Environmental Protection Agency specification i.e. 500 mg/l^{-1}.

The acidity of sample C was zero level but sample A and B shows some significant increase in acidity. In sample A and B, the acidity i.e. (10.50 mg/l^{-1} and 6.96 mg/l^{-1}) falls within the acceptable Federal Environmental Protection Agency standard.

The sulphate (site A=25.25, site B=66.926 and site C=175.02) mg/l^{-1} and chloride (site A=41.00, site B=111.14 and site C=120.40) mg/l^{-1} level of all the samples are within the acceptable standard of Federal Environmental Protection Agency, the highest of sulphate and chloride are 175.02 mg/l^{-1} and 120.40 mg/l^{-1} while Federal Environmental Protection Agency standard are 1000 mg/l^{-1} and 250 mg/l^{-1} respectively.

The total hardness of all the sample i.e. A=101.00, B=116.73 and C=141.73 all in mg/l^{-1}, complied with the Federal Environmental Protection Agency standard, the highest of all was 141.74 mg/l^{-1}, while Federal Environmental Protection

Agency standard is 500 mg/l^{-1}. Also the magnesium hardness (A=55.33, B=75.73 and C=53.33) mg/l^{-1}, the highest was 75.73 mg/l^{-1} while Federal Environmental Protection Agency standard was 200 mg/l^{-1} calcium hardness of sample C, is higher than Federal Environmental Protection Agency standard, i.e. 88.40 mg/l^{-1} while Federal Environmental Protection Agency standard is 75 mg/l^{-1} but sample A and B are within the Federal Environmental Protection Agency limit i.e. 46.66 mg/l^{-1} and 39.00 mg/l^{-1}.

The average concentration of lead in the sample are zero, the average concentration of iron (site A=0.756, site B=2.79 and site C=7.123) mg/l^{-1} in the sample, are within the acceptable standard except site C that was beyond FEPA standard of 5 mg/l^{-1}zinc shows a very high concentration in site B i.e. 10.076 mg/l^{-1} against less than 1.0 mg/l^{-1} specified by FEPA, but not detected in site A. the high concentration of these results could be the use of galvanized zinc as roofing sheet in most of the building in the area, which could be easily washed into the drainage when rain fall.

The iron concentration of samples were within the FEPA specification, the highest value of site A and site B was 2.79 mg/l^{-1}, only the site C gave a value higher than FEPA standard, i.e., 7.123 mg/l^{-1} against 5 mg/l^{-1} specified by FEPA.

Iron concentration may impact odour to the receiving water body and staining in laundry and in porcelain surfaces. FEPA specification while sample A and B i.e. 2.72 mg/l^{-1} 15.87 mg/l^{-1} are within. Sample C=112.083 mg/l^{-1} against 40 mg/l^{-1} maximum specified by FEPA, these results conformed to the research carried out by Lagos State Environmental Protection Agency in some pharmaceutical industries in Ikeja industrial area.

The average result of BOD was generally low in all the samples when compared with FEPA specification (A=14.33, B=18.176, C=12.84 and FEPA=50) ppm. Dissolved carbon (iv) oxide shows the lowest concentration in the sample A and sample C, 2.25 mg/l^{-1} and 12.850 mg/l^{-1} but high in sample B=45.34. High concentration of dissolved could be due to high organic or minerals acid in the effluent and this is corrosive to water handling equipment particularly steaming condensate equipment only small quality of carbon (iv) oxide is desirable as it helps to maintain the carbonate equilibrium, thus avoiding the formation of calcite scale on exposure to the surface.

The results of the microbial analysis did not meet the FEPA standard, except sample C i.e. colony forming per millilitre in all the plates are too numerous on the plate counts, coli form counts, mould and yeast while FEPA specified 350 cfu/ml for total plate count and nil for coli form count. These results show that there is a lot of decay of various food items that is responsible for the propagation of these microbes and this can adversely affect the environment by aggravating different types of disease and if care is not taken, it could also lead to environmental disaster.

Considering the parameter analyzed, sample C gave the highest values in the following pH, conductivity, alkalinity, total hardness, chemical oxygen demand and temperature, settle able solid and dissolved solids, i.e. 11.483 mg/l^{-1}, 5.73, mg/l^{-1} 135.16 ppm, 141.7 ppm, 112.083 ppm, 37.66°c, 0.4123 and 262.08.

Sample B gave the highest concentration only in magnesium (75.73 mg/l^{-1}) hardness while sample A, show the lowest contamination found in sample C could be due to the location of the site, considering the fact that this area comprises of textile, printing factory and mechanic village, all the mentioned industries above produce a lot of contaminants.

Generally, all the parameters analyzed in site A, are within the FEPA/WHO limit except the microbiological parameter did not meet FEPA/WHO specification, in C, acidity and micro-biological parameter did not meet FEPA/WHO specification, while site C pH, calcium, and chemical oxygen demand (11.48 ppm, 88.40 ppm and 112.09 ppm), were beyond the FEPA/WHO specifications. However, the above results show that the microbial analysis of the effluent is higher than FEPA specification. The colony forming per ml ($Cfu/mg/l^{-1}$) was too numerous to count in the entire sample except in sample C, also some parameters such as calcium, acidity, chemical oxygen demand, colour, pH and alkalinity are far higher than FEPA specifications this shows that most of the industries if not all in Isolo industrial environment has no pollution abatement programme for their effluent.

In all the sites analyses, site C sample gave the highest concentration of pollutants followed by site C, B, and lastly site A i.e. C > B > A.

Conclusively, FEPA (The Ministry of Environment) in conjunction with its state counterpart i.e. Lagos State Environmental Protection Agency (LASEPA) needs to enforce all the industries located in Isolo Industrial Estate to put in place an Effluent Treatment Plant (ETP), which would provide for physical, chemical and biological treatment of their effluent whose functions among other would include, precipitation of toxic substances and clarification with reduction in colloidal, biochemical oxygen demand and corresponding.

Pollution Control Measure

The harmful ecological and deleterious human health impacts of the dumping of the industrial effluent into water bodies have elicited National and International control measure to control pollution from these sources.

There are two basic types of prevention:

Preventive and Reactive Intervention

Preventive intervention can take varieties of forms, all with the objective of preventing the generation of the pollutants or their release into the environments. The objective of such approach is to protect an environmental media from contamination and as a consequence prevent organism or human being from exposure and thereby harm by toxic and hazardous substances.

Reactive Intervention is usually implemented to control the manufacture import use or release of substances that are known to be formed in manufacturing or industrial process or release into the environment as a result of accidents or spills. This control is refers to as "End of the pipe control" when they are

Comparison of Effluent Results with FEPA Standard Samples

Parameters	Site A	Site B	Site C	FEPA Standard
Colour	Off white	Off white	Dark Brown	Not stated
Odour	Slight	Slight	Pungent	Not stated
pH	6.67	6.266	11.483	6-9
Conductivity	0.203	0.726	5.73	Not stated
Alkalinity	No detail	No detail	1.352	500
Acidity	10.50	6.966	< 0.01	20
Sulphate	25.25	66.926	175.02	1000
Chloride	41.00	111.14	120.40	250
Total Hardness	101	116.73	141.73	500
Lead	0.01	0.443	0.143	< 1
Zinc	0.01	10.076	< 0.1	< 1
Magnesium	55.33	75.75	53.33	200
Calcium	46.66	39.00	88.40	75
Iron	0.756	2.79	7.123	< 5.0
C.O.D.	2.726	15.87	112.083	40
Dissolved Oxygen	21.66	31.80	17.01	Not stated
Dissolved CO_2	2.25	45.34	12.850	Not stated
B.O.D.	14.33	18.176	2.84	50
Temperature	29.66	28.00	37.66	40
Settle able solid	0.523	0.223	0.426	Not stated
Suspended Solid	0.523	0.0646	0.426	30 Maximum
Dissolved Solid	0.716	0.093	262.08	2000 Maximum
Flow Rate	0.06	0.066	0.1563	Not stated

Micro-biological Results

Total Plate Count (Cfu/ml^{-1})	Bacteria Too Numerous To Count	Bacteria Too Numerous To Count	Nil	350 Cfu/ml^{-1}
Coli form (Cfu/ ml^{-1}) (Pathogenic Bacteria)	Pathogenic & Bacteria Too Numerous To Count	Pathogenic & Bacteria Too Numerous To Count	Nil	150 MPN/100 n
Mould & yeast (Cfu/ ml^{-1})	Fungi Too Numerous To Count	Fungi Too Numerous To Count	Nil	Not stated

apply to by-product or emission and effluents from industrial operation. Some of the examples are indicated below:

- Restriction are apply to limit the concentration or qualities of the pollutants that may be presents as residue in the media or present in the mission or effluents releasè into the environments from industrial or others operation.
- Restriction may be placed on the limit of the pollutants that may be present in receiving environmental media.
- Requirement may be imposed to clean up site or media that has been contaminated as results of the previous into the environment.

End of the Pipe Treatment Technologies for Industrial Effluent

The objective of the end of the pipe treatment is to remove unwanted or ecological harmful or hazardous constituents of wastewater to render it harmless.

A wastewater treatment plant is also refers to as pollution abatement facility, Industrial building with an environmental concern recognized, especially in developed countries, would normally have waste treatment plant incorporated into industrials facilities so that the effluent quality can complied with statutory industrial effluent standard. The stories are different in developing countries such as Nigeria as indicated already in the foregoing until recently environmental consideration are lacking in the establishment of industries.

Consequently, less than 100 per cent have effluent treatment plant, even where pollution facilities abatement exists, and they are poorly maintained or are broken down or have capacity inadequate for volume of wastewater to be handled.

Pollution control required by FEPA and LASEPA now require the existing industries to reintroduce pollution abatement technology for compliance with minimum National standard for disposal of industrial effluent into the environment so as to protect the receiving water bodies and associated biota.

The use of both stream and effluent standard can be justifiable especially when there is a desire to adopt different standards base on industrial location, flow and quality of receiving water, climate and economic condition, for example, temporary adoption of both stream and effluent standard have being recommended for developing countries, because industrial pollution is limited to few urban centres while most of the surface waters in the rural area are clean with great assimilative capacity. Effluent standard in such area could be less stringent for a period of time.

A wastewater treatment facility may serve residential, commercial, and industrial septic tanks and public latrines. In developing countries, e.g., in Lagos State, conventional municipal wastewater treatment facilities generally are found only in connection with the sewered sections of large and well-developed metropolitan areas.

Raw sewage sludge and septic tank and latrine sludge should not be accepted for disposal at landfills because of the grave risk posed to human health, especially the landfill personnel. The grave risk is the consequence of handling and processing a material with the dense concentrations of human pathogens that are characteristic of such sludges.

Urgent Tips to Attain Best Practices

- All existing industries are to reintroduce pollution abatement technology in order to comply with National Standard.
- One or more comprehensive treatment plant, combining automatic separation and biochemical techniques must be established in all the industrial estate in Lagos State, with a daily disposal of at least 1500 tons.

Interrogating the Neo-institutional Thinking in Sustainable Development

14

Sam Wong

1. Introduction

Current calls for new institutional reforms in water management by the World Bank (2003), OECD (2004) and the European Union (2000) marks a significant concept difference from the previous ones. Underpinned by a new set of institutional thinking, the Sustainable Water Management Framework departs from the dualism between state ownership and full privatisation, and moves towards the partnerships amongst privatised water companies, public regulators and the community. Institutional strengthening by decentralisation, stakeholder participation and effective enforcement, as the framework suggests, can simultaneously achieve water efficiency, social equity and ecological sustainability (Dinar, 1998).

The new strategic water management is under the influence of neo-institutionalism. Institutions are not simply defined as regulations and organisations, but as 'rules of the game' (North, 1990). This broadens our understanding about the role of institutions in shaping water planning and distribution. The neo-institutional theory asserts that the failure of achieving sustainable use of water lies in inadequate incentives and weak regulatory arrangements. Institutional intervention, by purposefully crafting desirable forms of rules, regulations and coordination with water agencies, is an attempt to provide a long-term solution with low enforcement costs.

The framework has made significant contributions since it takes a holistic approach to water resources. It re-focuses on demand-side management (Griffiths, 2002). Deviating itself from neo-liberalism, this perspective acknowledges that water is both economic and public goods, and that the state and community should take a more proactive role in water governance.

There are, however, concerns about the universal application of the institutional design and crafting, without adequately addressing the contextual and cultural specificity. It is also accused of focusing on formal institutions and legal regulations and playing down the role of socially-embedded institutions (defined as social norms and values) in shaping people's perception about sustainable water use (Mehta et al., 1999).

The objective of this paper, therefore, is to explore the neo-institutional approach to water management and examine the effectiveness of the new institutional arrangements. It also aims to seek a better understanding of the role of institutions in water governance, and to uncover the latent tensions within the framework. England and Wales are chosen as a case study in this paper because they have experienced a dramatic change in water institutional structures, chronologically from state ownership to deregulation and then to current partnership of privatised water companies and public regulators.

The structure of this paper is as follows: it starts by introducing the sustainable water management framework by highlighting the key features, goals and assumptions of institutional reforms. It then gives a brief introduction to the institutional change in water management in England and Wales. The third section will examine the limitations of the institutional approach by arguing that institutions may create disincentives, rather than incentives, in achieving strategic water planning. Paradoxes within the sustainable management framework, such as the tension between the goal of efficiency and equity and between the interests of customers and shareholders, will also be explored.

2. Sustainable Water Management Framework

To understand the sustainable water management framework, it is critical to know what sustainable water management means. Although the term has been criticised for being over-used, it still lacks a clear definition that turns it into an elusive concept. The considerable coverage of water management in literature does not help a great deal since most of it does not offer a clear definition (e.g. Hunt, 2004). On some occasions, however, the concept is used in a very specific way. For example, Johnson and Handmer (2002) take the risk redistribution perspective and define sustainable water management as 'the management of the trade-offs between benefits on the one hand (among which is the risk reduction to humans), and costs on the other (among which is the risk of environmental damage)' (346).

Generally speaking, three key elements are commonly found in the literature of sustainable water management. They are: finance, governance and ownership. They are aimed at proper coordination and planning in order to ensure that water is distributed fairly, efficiently and effectively, and simultaneously to achieve economic, ecological and social dimensions of sustainability. In other words, SWM urges re-examination of 'the roles played by the state and other public and private actors' in water financing and governance (Figueres, et al., 2003: viii).

2.1 *Five elements of institutional reforms*

The sustainable water management framework is summarised in Diagram 1. The World Bank, the European Union and the International Water Association are the key players in promoting the new institutional reforms. The five elements in the framework are: privatisation, decentralisation, stakeholder participation,

Diagram 1: Sustainable water management framework

Elements	Neo-institutional thinking	Goals
1. deregulation, privatisation and user-pay principle 2. decentralisation 3. stakeholder participation 4. enforcement and monitoring 5. appropriate technology	* strengthening water governance by institutional reforms * restructuring incentives * getting institutions right by institutional crafting and design	* address market failure * resolve collective action dilemmas * achieve water efficiency, social equity in access to water and ecological sustainability

effective enforcement and monitoring, and appropriate application of technology.

Calls for deregulation and privatisation are the essential part of the water reforms. Deregulation re-defines water ownership and shows a reduction of government control in water governance. It is claimed that implementing privatisation by selling the previously state-owned water enterprises to private investors brings competition, and therefore better services. A new water pricing system and the principle of full recovery of costs of water services are considered to provide adequate incentives for efficient use of water (Kallis and Butler, 2001).

A decentralised approach to water regulation transfers the major responsibility to the market and the municipalities. Assertions are made that centralised practice involves high administrative costs and results in low efficiency. The World Bank suggests that: 'experience demonstrates that decentralized service delivery can break the vicious cycle whereby service level declines' (quoted in Dinar, 1998: 371).

Stakeholder participation represents a bottom-up approach in that the public are encouraged to participate in water management and decision-making. By information sharing, increasing transparency and open communication, different stakeholders feel the ownership of, and share the responsibility for, water management (Petresin, 2003).

Effective monitoring and enforcement mechanisms are also central to the framework. They lay down clear rules about what water-using behaviour is desirable. Sanctions are applied to 'bad apples' who fail to comply with the regulations. The role of technology in the framework is different from the old practice. While technical knowledge and water-saving innovations remain critical in providing reliable water delivery, more emphasis is placed on minimising the negative environmental impact caused by technologies.

2.2 *Neo-institutional thinking*

Neo-institutionalism is distinct from old institutionalism in two aspects: firstly, it takes a broader view of institutions. Institutions do not simply mean regulations or organisations, but 'rules of the game'. It focuses not merely on the constraining nature of institutions, but also the enabling side in shaping human action (Ostrom, 1992). Secondly, it does not conceive the state and private companies as necessarily mutually exclusive in water governance. It promotes the idea that they can work together closely as strategic partners. The public (to be precise, community) is also encouraged to take part in decision-making processes to feel responsibility for sustainable water management. The neo-institutional thinking reflects the changing perception about the nature of water as both economic and public goods.

The failure to achieve better water management, the neo-institutionalists claim, lies in inadequate incentives, weak governance and inefficient regulatory and legal arrangements. The market alone is unable to achieve both water efficiency and social equity at the same time because of the complex nature of water. The solutions, they propose, are to restructure water-related institutions in order to get the incentives right, to redefine water as economic goods, and to change the role and perception of the public and private sectors in the water industry. Success in attaining social equity in access to water, economic efficiency and ecological sustainability lies in having 'right' institutional arrangements. The right organisational structures help to identify rights and responsibility of different stakeholders. High level of public participation generates social capital and sense of ownership that facilitates water planning. Effective enforcement and monitoring is intended to deter free-riding behaviour, so as to align individuals' private interests with collective outcomes. These new institutional arrangements are regarded as a low-cost mechanism since they encourage negotiations and re-adjustment and secure a self-enforcing form of governance (Mansuri and Rao, 2003).

There are three forms of institutions: judiciary, bureaucratic and socially-embedded. Judiciary institutions comprise the law and regulations which determine the water ownership and structure of authority. Clear sets of rights and responsibilities are laid down, so that both water and sewage providers and water users know what is expected of them (El-Fadel et al., 2001). Sanctions, such as threatening to cut water supply, are used to deter free-riding behaviour and to ensure water efficiency. Bureaucratic institutions are water-use organisations and committees which make decisions about rules, access and distribution. In contrast, socially-embedded institutions are social values and norms which govern and shape our water-using behaviour. These three forms of institutions, however, are not mutually exclusive.

2.3 *Similarities and differences to neo-liberalism*

The neo-institutional approach marks a shift from the traditionally predominant supply-oriented modes of planning to the demand-side. Kallis and Bulter (2001) argue that:

> 'the **institutionalisation** of ecosystem-based objectives [are considered] as overriding criteria in water policy and decision-making. The [European Union] Directive **formalises** the environment as a "user" of water and establishes it at almost an equal footing with other human economic activities' (140, my emphasis).

The ideology of the framework is different from neo-liberalism. It promotes the 'private-public' synergy and takes a more positive view on the role of the community. It considers that the state can play a proactive role in creating a favourable environment for water governance. It also recognises that regulations are necessary to address market failure in achieving equitable water delivery. Environmental sustainability and water efficiency, from the neo-institutional perspective, are not necessarily a trade-off. By appropriate institutional designs, individuals and collective interests are aligned, and free-riding behaviour is deterred by the fear of legal and social sanctions. Despite the differences, the neo-institutional theory shares a few assumptions with the neo-liberal thinking: they both advocate a clearly-defined resource boundary and the specification of property rights.

3. Case Study: England and Wales

England and Wales have witnessed a dramatic historical transformation in water management over the past thirty years. Before the Conservative Party took power in 1979, the water industry was largely nationalised. Governments perceived that it was their duty to provide clean water to the general public. The privatisation scheme under the Conservative regime in the early 1980s, however, was influenced by a strong belief that market competition was the best way to promote water efficiency and to safeguard the interests of customers. Economic regulations, therefore, were kept at a minimum level at an interim stage of deregulation. The responsibility for environmental regulation was assumed to be transferred to the shareholders in the privatised companies (Summerton, 1998). To be fair, privatisation of water industry at that time was a response to economic recession and tremendous fiscal pressure. Research has shown that water industry had long suffered from under-investment and had poorly performed. The government, acting as both a regulator and service provider, had conflicting interests, and self-prosecution was very rare.

This minimal intervention, however, was severely challenged by consumer organisations and the media. Under the natural monopoly, water companies might seek profits at the expense of public interests. Pressure on the government to envisage adequate regulatory protection of consumers against the interests of service providers was mounting. In 1989, the government introduced a dual system: privatised companies were responsible for providing clean water and sewage treatment while the environmental and drinking water regulation was the responsibility of separate public bodies.

All these changes mark the influence of both domestic and international pressure. The changing socio-economic-political circumstances in England and Wales suggest economic arguments alone are unable to promote strategies for water institutional reforms. They require social legitimacy and political capacity. The Water Framework Directive by the European Union came into force in 2000. Griffiths (2002) suggests that the Directive serves as an incentive for enforcing joint international management, and 'in many respects, the UK already utilises the basic principles and philosophies set out in the Directive' (14). The paradigm shift about the understanding of the nature of water as both economic and public goods also plays a role in encouraging stronger regulation and legislation in water governance.

3.1 *Regulatory structure*

The regulatory framework functioning in England and Wales is summarised in Table 1. Ten water and sewage companies and 16 water companies are responsible for distributing and managing potable water supply and waste-water treatment. They are also required to design, implement and maintain the water infrastructure. Different public bodies oversee the water sector in different aspects. The Drinking Water Inspectorate (DWI) evaluates the quality of drinking water. The Office of Water Services (OFWAT) is an economic regulator which monitors the water price levels. The Environment Agency (EA) is responsible for monitoring the water environment.

Table 1: The division of labour in water provision and regulation

Authority	Duties
Water companies	• provide potable water to consumers and operate and maintain infrastructure facilities.
Drinking Water Inspectorate (DWI)	• evaluates the quality of drinking water.
Office of Water Services (OFWAT)	• monitors water price levels.
Environment Agency (EA)	• supervises waste treatment and prevents air and water pollution.

Sources: EA, OFWAT.

Experiences in England and Wales appear to suggest that successful water management lies in strategic partnership amongst private service providers, the government and different regulators (Kinnersley, 1998). The clear and simple institutional and regulatory arrangements provide adequate incentives for efficient governance. This managerial structure also tries to show that privatised water companies, under strong profit-seeking pressure, are not necessarily incompatible with long-term benefits for consumers and the environment. Whether these claims are valid in reality will be discussed in the subsequent section.

3.2 *Cultural and contextual influences*

England and Wales have implemented most of the policies suggested by the sustainable water management framework, such as privatisation, user-pay principle and involvement of public regulators. It is, however, interesting to recognise that they do not follow all the prescriptions. For example, the level of community participation in water governance in England and Wales remains low. My other paper (Wong, 2005) shows that water users in the developing countries are encouraged to participate in negotiating rules and roles. English and Welsh customers, though not completely neglected, are largely left out in the decision-making processes. For example, when Thames Water decides to build a desalination plant to increase water supply to meet the increasing demand, it is not certain how much the public are consulted or even informed about this decision and the choice of technology. The use of sanction is also seldom enforced in England and Wales. This is because of huge social pressure from the public and the media related to humanitarian and public health issues. Recent legislation also makes the disconnections of water supply for non-payment of bills illegal. Experiences in England and Wales, therefore, illustrate that sustainable water management is contextually specific and subject to cultural relevance. This casts doubts about the universal application of the neo-institutional approach to different places and cultures without critical scrutiny.

4. Limitations of the Neo-institutional Approach

An understanding of institutions is now seen as central to successful water management. However, how likely are these new institutional arrangements, established with a strong focus on privatisation, regulation, decentralisation and community participation, to provide the answer to sustainable water management in England and Wales?

Conceptually, the institutional perspective to strategic water management takes a 'take and mix' approach in institutional designs. New and efficient institutions are carefully moulded, while the old and inefficient ones are abandoned or forced to transform. It also shows an instrumental view on institutions.

Rules are purposefully crafted to produce desirable forms of collective action (Ostrom, 1992).

The problem of this approach is that it considers institutional building as a 'unilinear model of institutional evolutionism' (Cleaver, 2002) which assumes that new institutional reforms necessarily move towards higher efficiency. Another limitation is that the planned institutional interventions are designed to minimise uncertainty (Mehta et al., 1999). Petresin (2003), for example, argues that the strategic framework 'provides broad guidance for water management, sets out clear and strategic directions regarding how water should be managed or **result in specific actions**' (7, my emphasis). This approach, however, fails to recognise the complexity of the social and natural worlds and the unpredictable outcomes of human action.

4.1 *Institutions create disincentives*

OECD (2004) suggests the development and application of a strong legal framework to ensure the sustainable use of water. This is based on an assumption that regulations, stressing obligation and coercion, give the right signals for individual water users to align their private interests with collective outcomes. Regulations also guide water service providers to provide an appropriate level of service with an acceptable performance at an acceptable cost to customers (Entwistle, 1999: 384). Under a right institutional framework, institutional scholars believe that the conflicts between environmentalism and entrepreneurialism are reconcilable. In its consultation paper, CAG consultants, for example, suggest that: 'reforming the current regulatory regime could make it profitable for companies to provide better... services with less fuel, thus decoupling well-being from environmental damage' (2000, para 8.8). Faith in institutional crafting also highlights the positive role of institutions in reducing barriers to water conservation (Ward and King, 1998).

Evidence, however, shows a rather different picture—regulations hinder, rather than enhance, the adoption and implementation of sustainable practices. Foxon et al. (2002: 444) point out that: 'the imposition of institutional systems and regulatory targets will encourage the adoption of less sustainable technologies or solutions by the water industry'. In other words, the heavy reliance on laws and regulations in the water industry merely encourage seeking 'quick wins' rather than long-term, strategic and more innovative sustainable initiatives.

Disincentives arising from the regulatory framework, Willis et al. (2002) explain, are due to the profit-driven nature of private water companies. As their main objective is to guarantee a constant water supply to meet demand, they do not have a strong incentive to do more than meet current environmental regulations (Cashman et al., 2003). Berndtsson and Hyvönen (2002) also find out that it is the laws and regulations which obstruct the implementation of sustainable alternatives to water-based sanitation system. In their words, 'conventional water-based sanitation systems are fully legal, and even mandated by the legal system' (526).

In addition, standard regulations have difficulty in meeting inter-regional variations. Taking waste minimisation as an example, Phillips et al. (2000) mention that: 'although best value legislation is expected to make a marked contribution to the adoption of more sustainable waste management practices in the future, cognisance must be taken of the regional variation in MSW (municipal solid waste) minimisation programmes' (302). Owing to heterogeneity of different regions, treatment of environment issues should depend on the regional environmental resources.

4.2 *Playing down the role of socially-embedded institutions*

As discussed before, there are three forms of institutions: regulatory, bureaucratic and socially-embedded. In the sustainable water management framework, however, more emphasis is on the first two forms, but less on the third. Institutionalists tend to believe that formal institutions, such as laws, regulations and water organisations, are effective in ensuring effective use of water. It is also believed that 'inefficient' socially-embedded institutions are challenged and gradually be replaced by 'robust' formal institutions. The neglect of informal institutions, however, has negative impact on water sustainability. The dualism of formal and informal institutions also fails to see the interactive nature of institutional building.

Informal or socially-embedded institutions are defined as social values, norms, habits and routines guiding our everyday lives, including water consumption and perception of alternative water innovations. These institutions are the products of social and political practices. They are powerful in influencing our water-using behaviour because water-related norms and practices 'think on our behalf' (Douglas, 1987), and they can both facilitate and constrain individuals in switching to more sustainable innovations. For example, to explain the failure of the riparian buffer zones in England, Ducros and Watson (2002) interviewed the affected farmers about their goals, motives and values. Recognising the role of socially-embedded institutions explains why the technically sound policies can result in limited impact.

Decisions made by customers are not always confined by formal institutions. The water metering system, for example, is expected to be an effective demand management measure since it is intended to increase customers' sensitivity to water consumption and the costs. Evidence, however, shows that households in the North of England deliberately consumed more water during the drought in 1995 in order to protest against the mismanagement of the Yorkshire Water company (Haughton, 1998).

4.3 *Over-simplification of agency*

Providing adequate incentives is the central idea in the sustainable water management framework. This takes an economistic view of human behaviour, arguing that individuals take actions by constantly making the appropriate calculation of cost and benefit. This resonates with the rational choice theory, assuming that human beings are necessarily selfish and calculating, and thus they tend to cheat and abuse water. In order to avoid water wastage, heavy reliance on regulations and monitoring, such as metering, are in favour. Sanctions are considered as necessary to punish non-compliance. In the principle of water charging, Figueres et al. (2003) make this assumption explicit: 'paying for water makes people more responsible for their demand, as they **often ask for more than they really need'** (202, my emphasis). The case of BedZED (a new housing development in the London Borough of Sutton), however, demonstrates that, even without coercion, tenants are willing to cooperate and rarely put sanitary towels, nappies or condoms down the toilets or drains (Shirley-Smith, 2004). This example suggests that individuals can act altruistically and cooperatively for sustainable development.

Agency here is defined as individual's capacity to act. The tendency to reinforce the role of water users as 'customers' or 'consumers' is problematic since this has simplified the complexity of agency. Assigning clear rights and responsibilities to customers is considered as an act of empowerment which gives them a strong sense of ownership. Encouraging customer-orientation to develop financially viable water and sanitation services regards customers as rational individuals who make water-related decisions based on economic rationality. The cost-benefit analysis attempts to establish links between the kind of service that benefits the stakeholders and what they are willing to contribute in cash (Mulenga, 2002: 1).

The customer-orientation, however, does not necessarily guarantee higher transparency or a louder voice to water-users in the process of decision-making. The regional privatisation of the water market in England means that water companies have a monopoly of water distribution and sewage disposal within their franchise areas. Customers cannot vote with their feet by switching to another water utilities, no matter how dissatisfied they feel about the quality of water service. In addition, the main concern of water companies is to maintain financial returns and satisfy investors and shareholders. The needs of their customers are not really a very high priority. Empowering customers to ensure higher water efficiency and sustainability, therefore, remains a myth because customers lack the real power to challenge the status quo.

4.4 *Paradoxes within the sustainable management model*

There are latent tensions within the strategic water framework which needs to be addressed properly. The three levels of dilemmas are: the 'efficiency-equity' paradox, the conflicting interests between water customers and shareholders, and the tension between the demand- and supply-side management.

Petresin (2003) proclaims that: 'water must be managed and conserved in a fair and efficient manner' (6). This highlights that an equitable access to water is a social right. However, at the policy level, achieving water efficiency without undermining social equity needs a fine balance. Increasing water prices, for example, helps to improve efficiency, but it has negative impact on the livelihoods of the poor. Encouraging the public to participate in decision-making is also an attempt to resolve the 'efficiency-equity' paradox by instilling the sense of right of, and responsibility for, water conservation. Community participation, however, is not necessarily good for the poor since active participation demands extra time, resources and energy from the poor. Participation may be used merely as a means to improve project efficiency and to get funding from the government, rather than as an end itself to increase community ownership and commitment.

Some writers may take the view that the tension between shareholders and customers is not necessarily problematic. The obligation of the water companies to generate revenues for shareholders forces them to improve water efficiency which will then lower water prices and improve environment in the long run. In reality, however, water companies face tremendous commercial risks in business continuity. The pressure to meet the rising expectation of shareholders means that the larger the price cap, the greater the returns they can provide for shareholders (Johnson and Handmer, 2002). The profit-seeking image of water companies, portrayed by the media, also undermines trust between the public and the companies. When the services fall short of customers' expectations, it will easily trigger off the blame-game. During the drought in North England in 1995, for instance, the water companies accused the public of being 'culturally ignorant about the value of water' (Haughton, 1998).

The sustainable water management framework plays down the dimension of power. In England, improving drinking water quality and efficiency of water use still depends heavily on development in science and technology. It is not to deny the contribution of technical know-how towards sustainability, but the traditional 'predict and provide approach', devising engineering solutions to satisfying human needs, remains strong. The blue-print ideas about project planning and implementation still dominate the intervention and institutional reforms (Willis et al., 2002: 363). This atmosphere creates an unfavourable condition for community participation since local knowledge is not highly valued. The demand-side management would then easily be hijacked by the supply-oriented.

5. Conclusions

This paper has examined the currently influential neo-institutional approach to sustainable water management. The central idea of this approach is to build strategic partnership amongst private water companies, the government, public regulators and the community to achieve efficient water governance and financing. Policies such as decentralisation, community participation, effective monitoring and use of sanction, are all intended to restructure the incentive mechanisms, to craft the right forms of institutions, and to build a clear structure of authority that water companies and users can follow.

This paper does not deny the contributions that the neo-institutional thinkers have made in re-focusing the significance of institutions in shaping human action. Defining water as both economic and public goods helps to recognise that the market alone cannot achieve water efficiency, social equity in access to water and ecological sustainability. Drawing upon the community in the decision-making process and bringing the state back into play address some of the limitations of neo-liberal paradigm. As Johnson and Handmer (2002) stress: 'we need institutional arrangements that facilitate the management of, and anticipate the influence of, competing and conflicting demands' (346).

My research, however, has exposed the wide gap between the neo-institutional theory and current water realities in the case study of England and Wales. The discrepancy can be explained by the instrumental view on, and inadequate understanding of, institutions. Strong belief in the efficiency of new formal institutional arrangements by the active process of institutional design and crafting in pursuit of optimal resource management fails to consider the social impact on, and the cultural influence of, the public's perception of sustainable use of water. This paper has also showed that over-reliance on legal regulations can 'unwittingly' create disincentives. Downplaying the socially-embedded institutions undermines the effectiveness of bureaucratic institutions. The case study of England and Wales has also illustrated that implementing sustainable water management is contextually specific, and that the universal application of the design principles is problematic.

In addressing the shortcomings, it is suggested that institutional interventions in future should consider both the micro and macro process of institutional strengthening. Institutional building cannot be separated from people's belief and livelihoods. Micro contexts matter since they provide the conditions that shape water access and consumption on the one hand and willingness and capability to pay on the other. By far, little research has been done on how individuals respond to new water regulations. Qualitative studies then should complement quantitative research to investigate the effectiveness and desirability of sustainable water management.

The narrow focus of the sustainable water management framework on water aspects only fails to pay sufficient attention on a wider environmental concern. To alleviate the increasing housing demand, for instance, the government plans to build 50,000 new houses in drought-prone South East England. This illustrates that water sustainability gives way to economic growth. Only by integrating strategic water programmes into land and economic planning can the holistic approach of water management be genuinely achieved.

REFERENCES

Berndtsson, J. and Hyvönen, I. (2002). "Are there Sustainable Alternatives to Water-bases Sanitation System? Practical Illustrations and Policy Issues." *Water Policy*. 4: 515-30.

Cleaver F. (2002). "Reinventing Institutions: Bricolage and the Social Embeddedness of Natural Resource Management." *The European Journal of Development Research*, 14, 2: 11-30.

Dinar, A. (1998). "Water Policy Reforms: Information Needs and Implementation Obstacles." *Water Policy,* 1: 367-82.

Douglas, M., (1987). *How Institutions Think*. Routledge: London.

Ducros, C., Watson, N. (2002). "Integrated Land and Water Management in the United Kingdom: Narrowing the Implementation Gap." *Journal of Environmental Planning and Management,* 45, 3: 304-23.

El-Fadel, M., Zeinati, M., Jamali, D. (2001). "Water Resources Management in Lebanon: Institutional Capacity and Policy Options." *Water Policy,* 3: 425-48.

Entwistle, T. (1999). "Towards Sustainable Waste Management: Central Steering, Local Enabling or Autopoiesis?" *Policy and Politics,* 27, 3: 375-38.

European Union (2000). EU Water Framework Directive 2000/60/EU of the European Parliament and of the Council of 23 Oct., 2000.

Figueres, C., Tortajada, C. and Rockstrom, J. (eds). (2003). *Rethinking Water Management: Innovative Approach to Contemporary Issues.* Earthscan Publications Ltd., London.

Griffiths, M. (2002). *The European Water Framework Directive: an Approach to Integrated River Basin Management.* European Water Association.

Haughton, G. (1998). "Private Profits—Public Drought: the Creation of a Crisis in Water Management for West Yorkshire." *Transactions of the Institute of British Geographers,* 23, 4: 419-33.

Hunt, C.E. (2004). *Thirsty Planet: Strategies for Sustainable Water Management.* Zed: London.

Johnson, C. and Handmer, J. (2002). "Water Supply in England and Wales: Whose Responsibility is it When Things go Wrong?" *Water Policy.* 4: 345-66.

Kallis, G., Butler, D. (2001). "The EU Water Framework Directive: Measures and Implications." *Water Policy,* 3: 125-42.

Kinnersley, D. (1998). "Privatised Water Services in England and Wales: A Mixed Verdict after Nearly a Decade." *Water Policy,* 1: 67-71.

Mansuri, G., Rao, V. (2003). *Evaluating Community Driven Development: a Review of the Evidence.* The World Bank: Development Research Group, first draft (February).

Mehta, L., Leach, M., Newell, P., Scoones, I., Sivaramakrishnan, K., Way, S.A. (1999). "Exploring Understandings of Institutions and Uncertainty: New Directions in Natural Resource Management." IDS discussion paper 372, Institute of Development Studies: University of Sussex, UK.

Mulenga, M. (2003). "Impediments to the Implementation of Demand Responsive Methodology in Urban Sanitation Programmes in Zambia and South Africa." Paper Presented in The Alternative Water Forum, 1-2 May, UK: Bradford.

OECD (Organisation for Economic Cooperation and Development) (2004). *The OECD Environmental Strategy: Progress in Managing Water Resources*. www.oecd.org/env [15 Oct. 2004].

Ostrom, E. (1992). *Crafting Institutions for Self-governing Irrigation Systems*. ICS Press: San Francisco.

Petresin, E. (2003). *Slovenian Water Management and EU—Challenge and Opportunity*. European Water Association.

Phillips, P., Adams, K., Read, A., Green, A. (2000). "Regional Variations in Waste Minimisation in England: Challenges and Issues for Policy Development." *Policy Review Section*, 297-302.

Sharp, L. (2005). "Water Demand Management in the UK: Examining Value Choices." Paper Prepared for the Associated European Schools of Planning Conference, Vienna, July 2005.

Shirley-Smith, C. (2004). "Setting up Water Management for BedZED." Paper presented in Conference: 'Planning for water: a major challenge for the government's new sustainable communities', 21 October, London.

Summerton, N. (1998). "The British way in Water." *Water Policy*, 1: 45-65.

Ward, F., King, J. (1998). "Reducing Institutional Barriers to Water Conservation." *Water Policy*, 1: 411-20.

Willis, K.G., McMahon, P.L., Garrod, G.D., Powe, N.A. (2002). "Water Companies' Service Performance and Environmental Trade-offs." *Journal of Environmental Planning and Management*, 45, 3: 363-79.

Wong, S. (2006). "Sustainable Water—Lessons from the Developing World." *Engineering Sustainability*, 159: 55-62.

World Bank (2003). World Development Report: Sustainable Development in a Dynamic World—Transforming Institutions, Growth and Quality of Life. Washington DC: The World Bank.

Contributors

B. Anjan Kumar Prusty, Environmental Impact Assessment Division, Sálim Ali Center for Ornithology and Natural History, Anaikatty (PO), Coimbatore-641 108, India. E-mail: anjaneia@gmail.com

P.A. Azeez, is a Senior Principal Scientist and Head of the Environmental Impact Assessment Division at Sálim Ali Centre for Ornithology and Natural History (SACON), Coimbatore. He is basically an Environmental Chemist with a Ph.D from Jawaharlal Nehru University (JNU) of Delhi, and has research interest in related fields such as environmental management, sustainable development and environmental economics. He has published a number of scientific papers and reports on various environmental issues and has guided doctoral and postdoctoral researches. E-mail: azeezpa@gmail.com.

Biswajit Mandal, Department of Chemical Engineering, Haldia Institute of Technology, ICARE Complex, HIT Campus, Haldia, West Bengal, India.
E-mail: bmandal_1977@rediffmail.com

Mohammod Aktarul Islam Chowdhury, Senior most Faculty Member in the rank of Associate Professor and Head, Civil and Environmental Engineering Department, Shah Jalal University of Science and Technology, Sylhet. E-mail: aic_cee@yahoo.com

Simi T.B. is a Researcher with CUTS International and is a postgraduate in Law from School of Legal Studies, Cochin

University of Science and Technology, Kerala, India. Her area of interest includes Trade, Development and Environmental Issues. With a professional stint of about two years in CUTS International, Simi has experience in working on projects and writing research papers related to trade, environment and WTO issues from South Asian perspective. She is presently working on phase II of the project 'WTO Doha Round & South Asia—Linking Civil Society with Trade Negotiations' supported by Oxfam Novib, the Netherlands, which aims to establish a linkage between the civil society organisations and research institutions while conducting advocacy with the governments.
E-mail: stb@cuts.org/simi.balakrishnan@rediffmail.com

Nicholas Ndegwa Kimani, LL.B (Keele), LL.M (Lon). He is presently a doctoral candidate at the Regulatory Institutions Network (RegNet) of the Research School of Social Sciences (RSSS), Australian National University.
E-mail: Nicholas.Kimani@anu.edu.au

Bhabani Shankar Nayak is a Doctoral Research Fellow in Development Studies and working as a Tutor in Development Studies in the University's Centre for the Comparative Study of Culture, Development and the Environment (CDE), University of Sussex, UK. He has many publications attached to his academic research and teaching.
E-mail:b.s.nayak@sussex.ac.uk

Geeta Sinha has finished her Master in Gender and Development from the Institute of Development Studies (IDS) at the University of Sussex, UK. She has more than a decade wide experience of working in the grassroot development sector.

Satyapriya Rout, Lecturer, Department of Sociology, University of Hyderabad, Hyderabad-500046, A.P.
E-mail: sprss@uohyd.ernet.in

P.R. Arun has a doctorate from the Bharathiar University of Coimbatore and has got over ten years of research experience in various fields such as Ecology and Environmental Impact Assessment (EIA). Apart from EIA he has wide ranging research interests in various fields such as Ecological

entomology, insect seasonality, butterfly ecology and pharmaceutical pollution. He was associated with the Environmental Impact Assessment Division of Sálim Ali Centre for Ornithology and Natural History (SACON), Coimbatore, and has been involved in eighteen different Environmental Impact Assessment projects across the country on various developmental projects such as hydroelectric dams, long distance gas/petroleum pipelines, bauxite mines, Thermal power plants, etc. He had also worked as Senior Officer at Environmental Justice Initiative, Mumbai. He has several publications to his credit in various reputed national and international journals as well as in popular media. Presently he is working with 'Green Alternatives' of Mumbai, as a consultant and Environmental expert.
E-mail: prarun@gmail.com

V. Mahajan Maya is a botanist having her Ph.D in Environmental Sciences. She was associated with various environmental research projects and her area of work includes weed ecology, environmental quality monitoring, fair trade, plant taxonomy, mangrove ecology and education. She is presently associated with the SPG Centre for Marine Ecology of Vikhroli, Mumbai.
E-mail: mayaruchi@yahoo.com

N.R. Vimala, worked for more than 20 years in the area of planning in CSIR. She obtained her MA in economics from University of Delhi, M.Sc degree in National Development and Project Planning from University of Bradford, UK and Ph.D in Policy and Planning from BITS Pilani. Dr. Vimala's areas of research interest are sustainable development, policy, planning and implementation. She has presented papers in conferences on sustainable development in the areas of water resources, leather, from concept to implementation.
E-mail: vimala_nr@yahoo.com

Agnes Atia Apusigah is a political sociologist with interest in gender issues, development criticism, educational reform and participatory approaches. She is a lecturer in African Studies, Gender and Development Studies and Development Theory in the University for Development Studies, Ghana.

She has co-edited a number of texts including *African Knowledges and Sciences* (with D. Millar, S.B. Kendie & B. Haverkort), *Action Research* (with D. Millar), *Participatory Training Techniques Using PRA Tools* (with D. Millar). She has also published papers including Diagnosing Poverty in Northern Ghana; Building Contingent Foundations; Gender Mainstreaming the Ghana Poverty Reduction Strategy or is it?; Two Illusions, One Reality; Performativity, Identity and Sexuality among Gurusi Women of Northern Ghana and; Meeting Women's Strategic Needs through Micro-enterprise Development. She is the editor-in-chief of the *Ghana Journal of Development Studies* (GJDS) and the *Journal for Educational Development in Africa* (JERDA).
E-mail: apusigah59@yahoo.com

Govindasamy Agoramoorthy, Department of Pharmacy, Tajen University, Yanpu, Pingtung 907, Taiwan.

Minna J. Hsu, Department of Biological Sciences, National Sun Yat-sen University, Kaohsiung 804, Taiwan.

Adewole Taiwo Adegboyega is an environmentalist. He is the founder of Environmental Support Services, Lagos Nigeria. He finished his Master Degree in Environmental Resources Management in 2004 from Lagos State University, Nigeria. He is also doing his Ph.D in the field of Environmental Management. He has written a lot of environmental related topics. He is a member of many professional bodies both in Nigeria and UK.
E-mail: environmental_ss@yahoo.com

Sam Wong (Ph.D), Lecturer, School of Earth and Environment, University of Leeds, UK.
E-mail: s.wong@see.leeds.ac.uk, kfswong@yahoo.co.uk